Genetics

A Guide for Students and Practitioners of Nursing and Health Care

HEALTH AND SOCIAL CARE TITLES AVAILABLE FROM LANTERN PUBLISHING LTD

ISBN	Title
9781906052041	Clinical Skills for Student Nurses
9781906052065	Communication and Interpersonal Skills
9781906052027	Effective Management in Long-term Care Organisations
9781906052140	Essential Study Skills for Health and Social Care
9781906052171	First Health and Social Care
9781906052102	Fundamentals of Diagnostic Imaging
9781906052133	Fundamentals of Nursing Care
9781906052119	Improving Students' Motivation to Study
9781906052188	Interpersonal Skills for the People Professions
9781906052096	Neonatal Care
9781906052072	Numeracy, Clinical Calculations and Basic Statistics
9781906052164	Palliative Care
9781906052201	Professional Practice in Public Health
9781906052089	Safe & Clean Care
9781906052157	The Care and Wellbeing of Older People
9781906052225	The Care Process
9781906052218	Understanding and Helping People in Crisis
9781906052010	Understanding Research and Evidence-Based Practice
9781906052058	Values for Care Practice

9781908625007

9781908625014

9781908625021

9781908625175

9781908625144

REVISED EDITION

Genetics

A Guide for Students and Practitioners of Nursing and Health Care

Karen Vipond

Lecturer, Bangor University

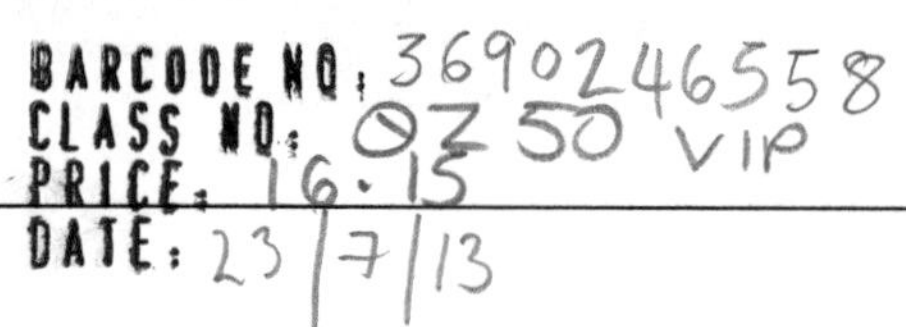

ISBN: 978 1 908625 15 1
Revised edition published in 2013 by Lantern Publishing Limited

First edition (ISBN 978 1 906052 24 9) published in 2011 by Reflect Press Limited

Lantern Publishing Limited, The Old Hayloft, Vantage Business Park, Bloxham Rd, Banbury, OX16 9UX, UK

www.lanternpublishing.com

British Library Cataloguing in Publication Data
A catalogue record for this book is available from the British Library

The authors and publisher have made every attempt to ensure the content of this book is up to date and accurate. However, healthcare knowledge and information is changing all the time so the reader is advised to double-check any information in this text on drug usage, treatment procedures, the use of equipment, etc. to confirm that it complies with the latest safety recommendations, standards of practice and legislation, as well as local Trust policies and procedures. Students are advised to check with their tutor and/or mentor before carrying out any of the procedures in this textbook.

Typeset by Medlar Publishing Solutions, India
Cover design by Andrew Magee Design Ltd
Printed and bound by TJ International Ltd, Padstow
Distributed by NBN International, 10 Thornbury Rd, Plymouth PL6 7PP, UK

CONTENTS

FOREWORD

For many, both public and health professionals, the world of genetics is surrounded by science, white coats and laboratories. For some it is associated with genetic modification, Mutant Ninja Turtles and 'designer' babies. Most books about genetics are written by experts, for experts and these can be a difficult read for anyone not an expert. What a delight it is therefore to read a book which takes complex scientific concepts and ensures they are understandable by all.

This book takes a step by step journey from genetic science to health via rare single gene conditions and conditions with both a genetic and environmental influence.

As a nurse and a nursing lecturer for many years, genetics did not seem to be a subject I needed to know anything about. We now know that a person's individual genetic make-up and the genetics of their disease can have implications for diagnosis, outcomes and treatment decisions. It is on this basis that genetics and genomics now need to be considered a core subject for all nurses and health care students.

I am pleased to recommend this book to students of nursing and health care. I feel the book also has merit for many practitioners who, like me, did not have the topic of genetics included in their training, and it will be a very useful book for health care educators in planning the content of future courses.

I congratulate Karen on producing a book that makes sense of the science of genetics and genomics and I hope you enjoy learning more about this exciting subject which is having such an impact on health care.

Candy Cooley, Genetics Awareness Programme Lead,
NHS National Genetics Education Centre

January 2013

INTRODUCTION

BACKGROUND

Following the publication of the government's White Paper 'Our Inheritance, Our Future' in 2003, the NHS Genetics Education and Development Centre was created. The Centre's work has included the publication of genetic workforce competencies that apply to all non-genetic health care professionals. Genetic education now forms an essential part of pre-registration nursing and midwifery courses so that all students can demonstrate competency in genetic practice at the point of registration.

This book has developed from the need for a genetic text applicable for health care practice. Common feedback obtained from health care students in the past has identified the lack of an available suitable genetic text. Many students have resorted to texts designed and written for medical students or for biology undergraduates. Most of the students felt that sections of these texts were either too in-depth or irrelevant to health care practice and therefore wanted a text that was relevant to them. This book was written primarily for student nurses, but the information within the book is also relevant for all other health care students, as well as for qualified professionals who would like to brush up on their knowledge of genetics.

The book is set out in ten chapters, starting with the very basics of cell biology. It has been designed so that each chapter builds upon the information given in the previous chapters. However, this does not mean that you have to start at Chapter 1; the chapter you start with will depend upon your prior knowledge and understanding of biology and genetics. Some readers might want to use the book as a reference source for genetic disorders; others might want to use it as supporting material for their health care course.

BOOK STRUCTURE

All the chapters have been written as clearly and concisely as possible in order to actively support learning. The book is organised as follows.

Chapter 1: Basic Cell Biology

This chapter provides a basic overview of the functions of the cell in relation to protein synthesis. Chromosomal structure is examined together with the basic units of inheritance, the genes. This is addressed to individuals who have no previous knowledge of cell biology.

Chapter 2: Inheritance

The Mendelian principles of transmission are explained in this chapter by using common genetic traits as examples. Possible allele combinations in offspring are demonstrated through the use of Punnet squares. Exceptions to the Mendelian rules are also outlined in this chapter.

Chapter 3: Autosomal Recessive and Dominant Inheritance

The inheritance of genes that are situated on the 22 paired autosomes are explained in this chapter, including examples of conditions that affect health. Transmission of recessive and dominant genes is explained, together with the classification of gene action. Incomplete dominance and co-dominance are outlined as well as the action of lethal alleles.

Chapter 4: Sex-linked Inheritance

The structure and the inheritance patterns of the X and Y chromosomes are explained in this chapter. Examples of genetic disorders arising from genes carried on the sex chromosomes are provided. X-inactivation through Barr bodies is explained as well as the process of Lyonisation. Sex-limiting and sex-influenced alleles are also explained in this chapter.

Chapter 5: Two or More Genes

This chapter covers both monogenic and polygenic inheritance. The calculation of probability of two or more monogenic traits is explained as well as the additional effects of polygenic traits. Multifactorial inheritance of common disorders such as diabetes, mental health problems and cardiovascular disease is explained, together with the heritability and empiric risk measurements of inheriting a multifactorial condition.

Chapter 6: Mutations

Chromosomal abnormalities and gene alterations are explained in this chapter. The different classifications of mutations, together with examples of clinical genetic conditions, are provided.

Chapter 7: Pedigree Analysis

The ability to take a family's medical history is essential to detect the mode of inheritance of a genetic condition and to estimate an individual's risk of developing that condition. This chapter explains the way in which pedigree charts are constructed, together with the recognised symbols, and opportunities are provided to practise the drawing of pedigree charts.

Chapter 8: Clinical Applications

Genetic screening, testing and gene therapy are explained in this chapter. Advances in technology have led to significant changes in health care, especially with regard to pharmacogenetics or 'personalised medicine.' Although many advances are still at the clinical trial stages, these are also explained in this chapter.

Chapter 9: Cancer Genetics

There has been a rapid development in the knowledge and understanding of genetics relating to different forms of cancer over the past decade, especially in relation to risk assessments, pharmacogenetics and disease prevention. Knowledge of cancer genetics has now become essential for all areas of health care and not just for those working in oncology units. The basics of cancer genetics are explained in this chapter, together with examples of some common cancers.

Chapter 10: Genetic Counselling

This chapter covers the specialist genetic services on offer to patients. Issues such as the aims of the service, appropriate referrals and the ethical issues of genetic testing are outlined. Access to accurate and relevant information is also covered, given that there is so much information and misinformation available on the Internet.

ACTIVITIES AND COMPETENCIES

There are plenty of activities included within the book for you to test your understanding before moving on to the next topic. Most importantly, answers to the activities are supplied at the back of the book. Most health care professionals are constantly involved in the caring for people with genetic conditions, and this book will provide you with genetic knowledge so that you can achieve the workforce competencies while you are working with patients.

The information provided in different chapters relates to different workforce competencies, but Chapters 1 and 2 form the basic foundation knowledge of the science of genetics.

Table 1 *UK Workforce Competencies for Genetics in Clinical Practice for Non-Genetics Health Care Staff (National Genetics Education and Development Centre, 2007)*

Workforce Competency	Relevant Chapter
1. Identify where genetics is relevant in your area of practice	3, 4, 5, 6, 7, 8, 9, 10
2. Identify individuals with or at risk of genetic conditions	3, 4, 5, 6, 7, 8, 9, 10
3. Gather multi-generational family history information	7
4. Use multi-generational family history information to draw a pedigree	7
5. Recognise a mode of inheritance in a family	3, 4, 5
6. Assess genetic risk	2, 3, 4, 5
7. Refer individuals to specialist sources of assistance in meeting their health care needs	10
8. Order a genetic laboratory test	8, 10
9. Communicate genetic information to individuals, families and other health care staff	10

AUTHOR BIOGRAPHY

Karen Vipond is a lecturer at the School of Healthcare Sciences at Bangor University, UK. She has combined her different careers as a biologist and a nurse in order to teach biological sciences to health care professionals. Karen's career experiences have been quite varied as she has worked as a biologist, a teacher, a trauma nurse, a district nurse, a health visitor and a medical research co-ordinator for Oxford University and the World Health Organization. Karen was a student at St Anne's College at the University of Oxford as well as Oxford Brookes University and she trained as a nurse at the John Radcliffe Hospital, Oxford. During her clinical career she worked in Oxfordshire, Buckinghamshire and in North Wales.

ACKNOWLEDGEMENTS

I would like to thank the following individuals for their help in the production of this book:The staff and students at the School of Healthcare Sciences at Bangor University for acting as 'models' for the photographs in Chapter 2; Ifor Williams, IT Technician at Bangor University for taking the photographs in Chapter 2; Marion Poulton, Librarian at the School of Healthcare Sciences for her help with proofreading the work; Carolyn Owen, Genetic Counsellor at Ysbyty Gwynedd, Bangor, for being a critical reader for Chapter 10; Joshua Vipond for his work in checking the scientific explanations of inheritance throughout the book; Judith Harvey for all her help and advice during the whole process of writing the book.

Copyright material

I would also like to thank the National Library of Medicine – National Institutes of Health, Bethesda, Maryland, USA, for their permission to reproduce illustrations in this book.

Dedication

To Joshua and Libby who have inherited half my genes, and to Phil who also contributed half his genes. Also to the rest of my family with whom I share the same genetic material.

01

BASIC CELL BIOLOGY

LEARNING OUTCOMES

The following topics are covered in this chapter:

- Cellular structure and function;
- Chromosomes;
- The cell cycle;
- Deoxyribonucleic acid (DNA);
- Protein synthesis;
- Mitochondrial DNA.

INTRODUCTION

The activities that occur within cells give us an understanding of how human traits are inherited. Knowledge of cellular function gives rise to the understanding of how the body works. The human body is made up of trillions of cells, many of which have specialised functions. Despite this, all cells share certain features:

- cells arise from the division of pre-existing cells;
- cells interact, they send and receive information;
- cells produce proteins for growth repair and normal body functioning;
- cells contain all the genetic instructions for the body.

All cells in the body behave in this way apart from red blood cells. Red blood cells are not considered to be true cells by the time they reach the blood stream as they do not contain a nucleus. Cells are the basic building blocks of all living matter.

CELL STRUCTURE

Cells have many parts, each with a specialised function. Any structure within the cell that has a characteristic shape and function is termed an **organelle**. Most organelles are too small to be seen through a light microscope but can be seen with an electron microscope (see Figure 1.1).

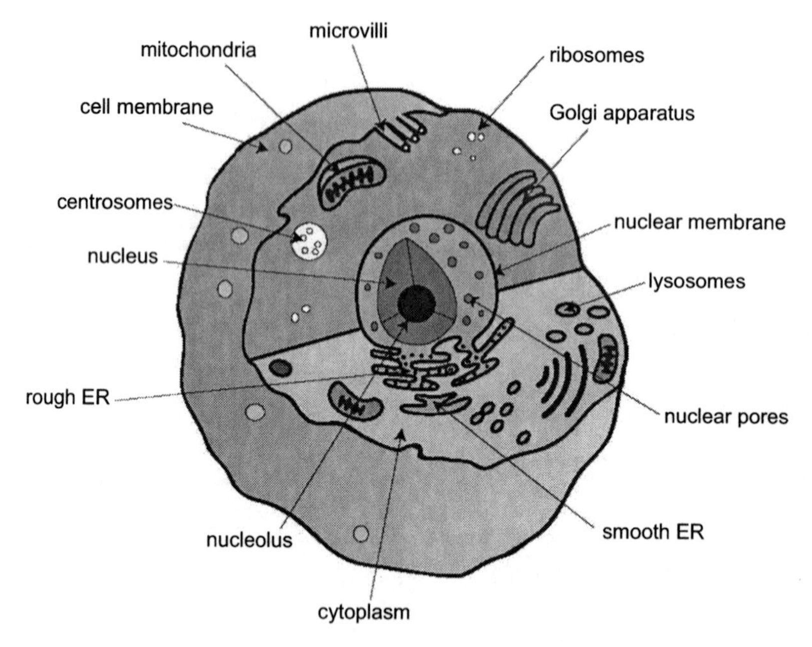

Figure 1.1 *Cell structure*

Plasma membrane

This is the outer lining of the cell. It is composed of a bilipid layer through which certain molecules can enter the cell (endocytosis) and wastes can exit (exocytosis).

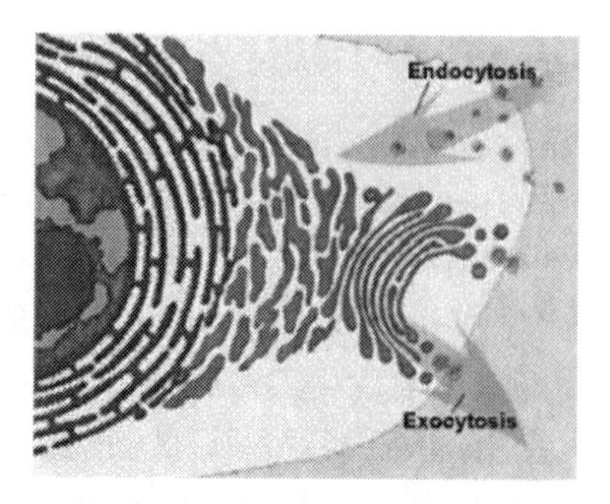

Figure 1.2 *The outer lining of the cell*

Nucleus

The nucleus functions as the control centre of the cell (Figure 1.3). It contains DNA (Deoxyribonucleic Acid) which is the cell's genetic material. A double membrane separates the contents of the nucleus from the rest of the cell. This nuclear membrane (also called the nuclear envelope) is perforated by nuclear pores.

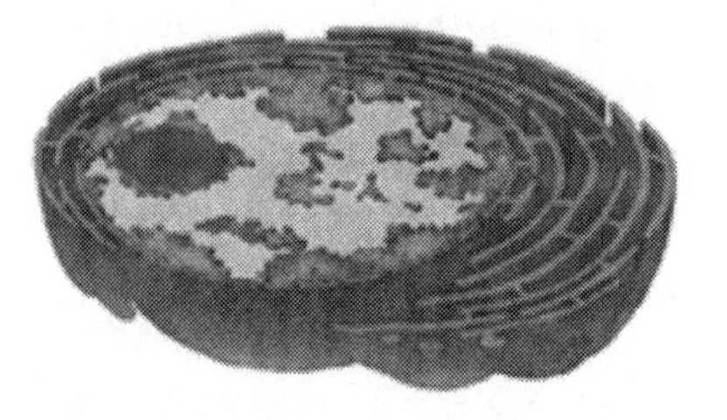

Figure 1.3 *Nucleus*

Nucleolus

The nucleolus (Figure 1.4) is a morphologically distinct area within the nucleus which is involved in the production of Ribonucleic Acid (RNA).

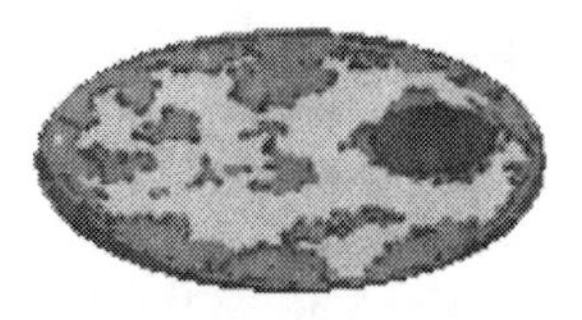

Figure 1.4 *Nucleolus*

Cytoplasm

Cytoplasm is a gel-like fluid that contains all the organelles and the enzymatic systems which provide energy for the cell.

Cytoskeleton

The cytoskeleton is a network of fibres made from the protein tubulin (Figure 1.5). This provides the structural framework of the cell and functions in cellular shape, cell division and cell motility, as well as directing movement of the organelles within the cell.

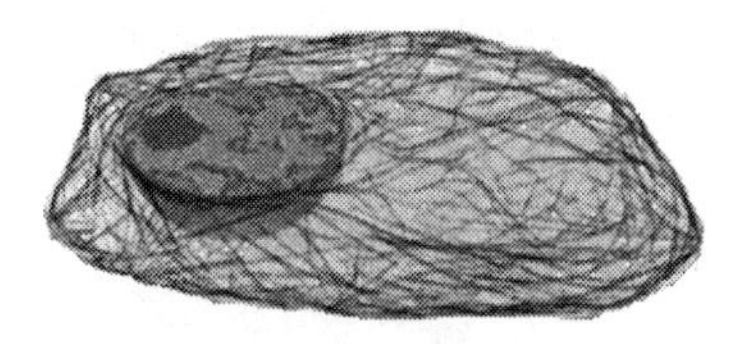

Figure 1.5 *Cytoskeleton*

Endoplasmic reticulum

The endoplasmic reticulum is an organelle that processes the molecules made by the cell (Figure 1.6). The endoplasmic reticulum transports these molecules to their specific destinations.

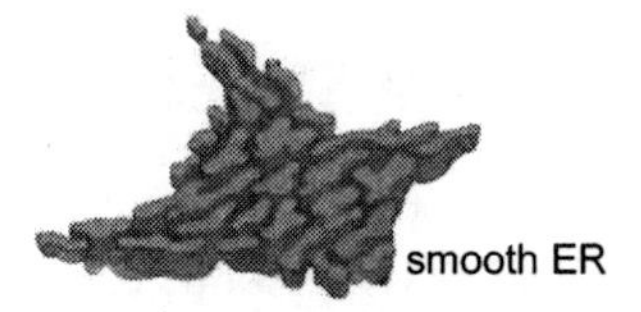

Figure 1.6 *Endoplasmic reticulum*

Ribosomes

Ribosomes are organelles that provide the sites for protein synthesis (Figure 1.7). Ribosomes are attached to the endoplasmic reticulum as well as freely floating in the cytoplasm.

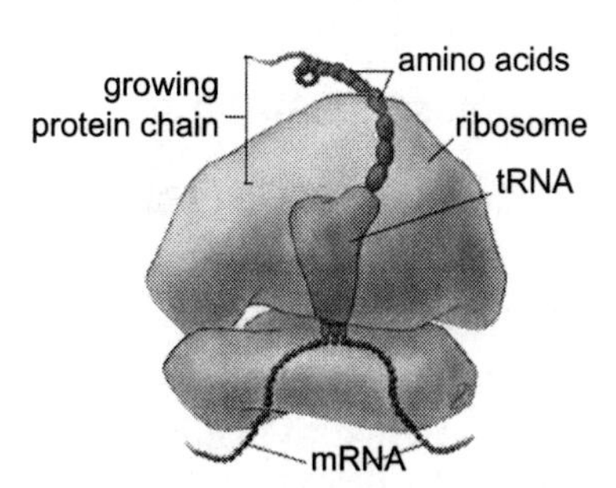

Figure 1.7 *Ribosomes*

Golgi body

The Golgi body is a structure that packages the molecules produced by the endoplasmic reticulum ready for transport out of the cell (Figure 1.8).

Figure 1.8 *Golgi body*

Mitochondria

Mitochondria are organelles that convert energy gained from food into a form that the cell can use (Figure 1.9). Adenosine triphosphate (ATP) is the main source of energy used by the cell. These organelles have their own genetic material and can make copies of themselves.

Figure 1.9 *Mitochondria*

Lysosomes

Lysosomes are organelles that break down bacteria and other foreign bodies, as well as recycling worn out cell components (Figure 1.10).

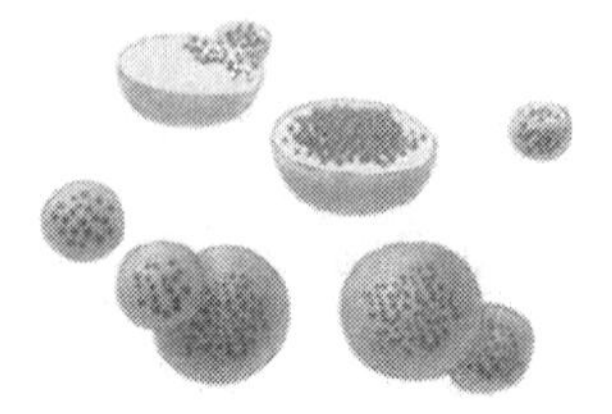

Figure 1.10 *Lysosomes*

Peroxisomes

Peroxisomes are responsible for the detoxification of foreign compounds and the oxidation of fatty acids (Figure 1.11).

Figure 1.11 *Peroxisomes*

CHROMOSOMES

Each of the trillions of cells in the body, with the exception of red blood cells, has a nucleus. Within each nucleus are structures called chromosomes. Chromosomes are not usually visible under a light microscope, but when a cell is about to divide, the chromosomes become denser and can be viewed at this stage.

Chromosome structure

A chromosome is composed of DNA and proteins and includes structures that enable it to replicate and remain intact (see Figure 1.12). During cell division, chromosomes have a constriction point termed a **centromere**. The centromere divides each chromosome into two sections or 'arms'. The long arm is referred to as the q arm and the short arm as the p arm (p for petite).

The location of the centromere gives the chromosome its characteristic shape and can be used to describe the location of specific genes.

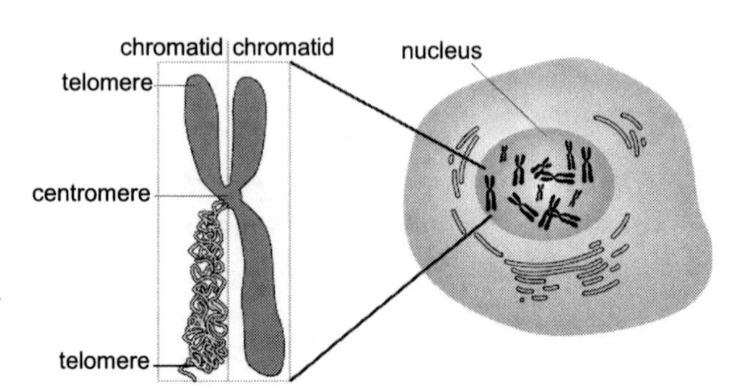

Figure 1.12 *A chromosome*

Telomeres

Telomeres are distinctive structures found on the end of each arm of the chromosome. They are made up of the same short sequence of DNA, which is replicated about three thousand times. The function of the telomeres appears to be twofold.

1. They protect the chromosome by 'capping' off the ends to prevent them from sticking or joining onto other chromosomes.
2. Due to the way that chromosomes are replicated, the ends of the chromosomes are not copied. Telomeres shorten during every cell replication, but the loss of DNA within the telomeres protects against loss of essential DNA within the chromosome itself.

Chromosome numbers

Chromosomes exist in pairs. Although not actually joined together, each pair has a characteristic length. The human cell nucleus has 23 pairs of chromosomes; in other words, 46 individual chromosomes. One chromosome from each pair is inherited from the father and one from the mother. Twenty-three individual chromosomes are inherited from each parent. The total number of chromosomes in each cell is called the **diploid number** (diploid 46) while the number of pairs is called the **haploid number** (haploid 23).

Of the 23 pairs of chromosomes, 22 pairs are termed **autosomes** and do not differ between the sexes. For ease of identification, these autosomes are numbered from 1 to 22. The chromosomes are numbered according to length, with chromosome number 1 being the longest and chromosome 22 being the shortest. The remaining two chromosomes are known as the **sex chromosomes**. These two chromosomes are not numbered but are known as the **X chromosome** and the **Y chromosome**. The Y chromosome determines maleness. A female will have two X chromosomes while a male will have one X and one Y chromosome.

Karyotype

The chromosome complement within the nucleus is called a **karyotype**. Charts called **karyographs** (see Figure 1.13) display chromosomes in pairs in size order. The 22 paired autosome chromosomes are displayed first, ranging from number 1 to 22 (largest to the smallest). The sex chromosomes, X and Y (male) or X and X (female) are always placed at the end of the chart. Karyographs can be a useful clinical tool to help confirm diagnosis through the identification of chromosomal aberrations, abnormalities or anomalies.

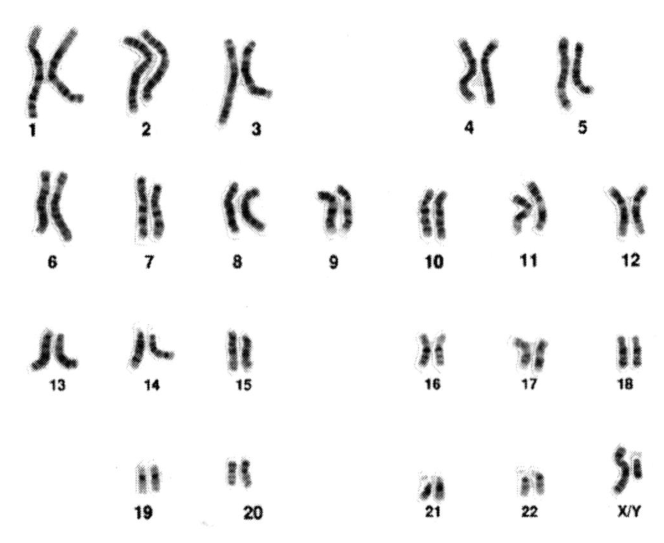

Figure 1.13 *A karyograph*

The centromere

Another physical characteristic of the chromosome, the centromere, also helps identification, as the position of the centromere varies in different chromosomes (see Figure 1.14).

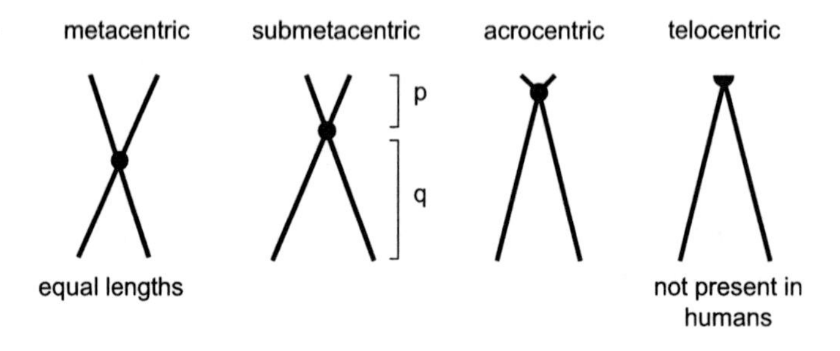

Figure 1.14 *Centromere positions*

ACTIVITY 1.1

What are the haploid and diploid numbers of chromosomes in humans?

CHROMOSOMAL INHERITANCE

The human cell has two sets of chromosomes, one set inherited from each parent. The complete genetic makeup within the cell is termed the **genome**. The total number of chromosomes within the cell has to be kept constant from one generation to the next. Each individual has a total of 46 chromosomes in each cell nucleus, 23 of which are inherited from their mother and 23 from their father.

For normal cell division two daughter cells are formed, both of which have the full 46 chromosome complement. This type of cell division is called **mitosis** and results in new cells that are genetically identical to the parent cell. Mitosis is cell division that is used by the body for growth and repair. **Meiosis**, on the other hand, is a type of cell division that produces new cells with only half the chromosomal complement (a total of 23 chromosomes). These 23 chromosomes are half the set of the original cell. Meiosis only occurs in the **germ line** cells, i.e. the ova in women and the sperm in men. If fertilisation occurs, the resulting offspring will inherit 23 chromosomes from the mother and 23 chromosomes from the father, resulting in a full 46 chromosomal complement. Meiotic division prevents the doubling of chromosomal numbers from one generation to the next.

Mitosis

Mitosis occurs rapidly during growth and tissue repair. It is a well-controlled process and consists of two major steps – the division of the nucleus followed by the division of the cytoplasm. Although mitosis is a continuous process it can be described as a series of four stages followed by a resting period where there is no cellular division (Table 1.1).

Table 1.1 *The stages of mitosis*

Stages	Events
Prophase	Chromosomes get shorter and fatter by coiling themselves. They now become visible under a light microscope. Each chromosome has two strands (two copies of the original chromosome) that are held together by the centromere. Strands of protein called spindle fibres appear.
Metaphase	Chromosomes line up together and the spindle fibres become attached to each side of the centromere.
Anaphase	The spindle fibres contract, pulling the two copies of each chromosome to opposite areas within the nucleus.

Telophase	The two new sets of chromosomes form two new nuclei. The chromosomes revert to being long and thin. The cytoplasm then divides to form two new cells.
Interphase	Normal cellular function. The cells make copies of their chromosomes ready for the cycle to start again.

With mitosis each daughter cell is an exact copy of the previous cell. All cells receive identical chromosomal material.

The cycle of events during mitosis usually lasts several hours. The mitotic division of the chromosomal material during prophase, metaphase, anaphase and telophase takes a relatively short period of time and the resting phase (interphase) takes up most of the time within the cell cycle (see Figure 1.15).

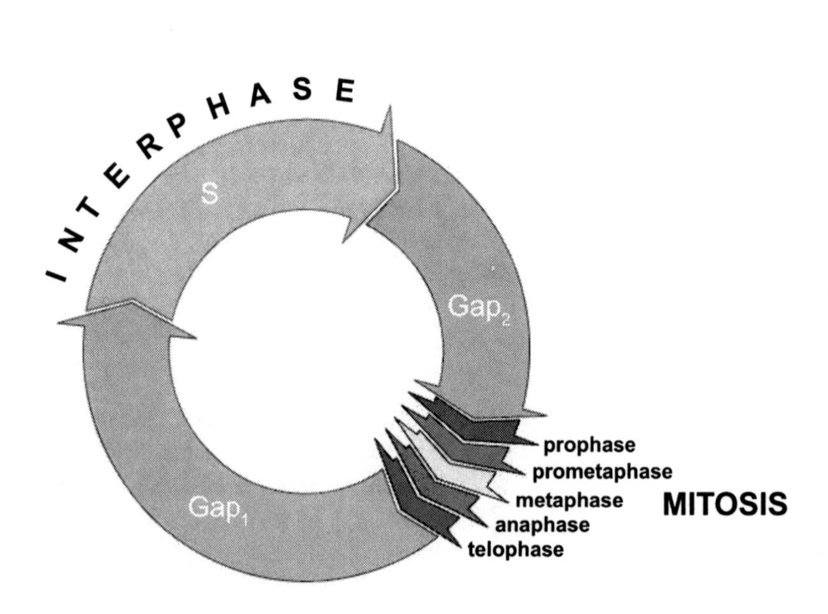

Figure 1.15 *The cell cycle*

The whole cell cycle takes approximately 24 hours, although this depends on which type of cell is involved. Mitosis usually only accounts for about an hour. Interphase is when no cellular division takes place. However, even during interphase, the cell needs to get ready for division so it increases in size. This stage is known as Gap 2 or G2. After division the cell needs to continue to grow so that it can achieve its optimum size; this is known as Gap 1 or G1.

Normally cells can undergo a total of 80 mitotic divisions before the cell dies, although this is dependent on the age of the individual.

Meiosis

Each cell contains two sets of chromosomes which exist in pairs. Meiosis results in cell division that produces new cells with only half the chromosomal complement. This is needed for the formation of germ cells (sperm in men, ova in women) so that two germ cells can fuse to form a full chromosomal complement.

Halving the full complement is achieved in two steps called meiosis I and meiosis II. Meiosis I is very similar to mitotic division in that two daughter cells are produced, both with 46 chromosomes. The main difference is that meiosis I takes much longer in comparison to mitosis and results in the 'crossing over' of chromosomal material (see Figure 1.16). Chromosomes 'swap' or exchange pieces of their structure with their partner chromosome before separating. This results in the daughter cells not having identical genetic material. This is the cause of genetic variability between individuals.

Meiosis II does not involve chromosomal replication but does involve the stages of prophase, metaphase, anaphase and telophase where chromosomes separate, new nuclei are formed and the cell splits into two. At the end of meiosis II the cells contain 23 individual chromosomes.

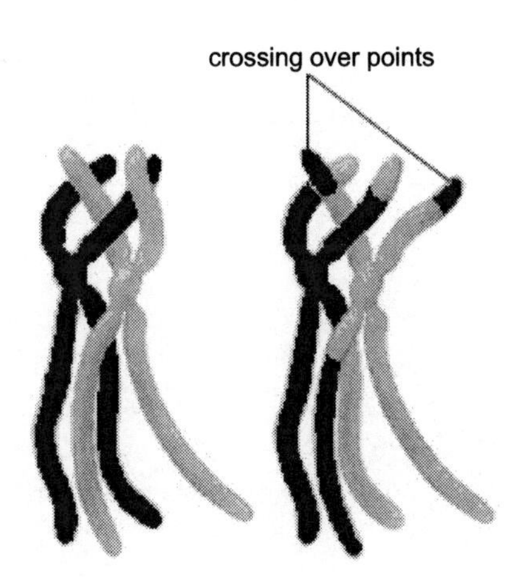

Figure 1.16 *Crossing over*

Differences between mitosis and meiosis

The differences between mitosis and meiosis are illustrated in Figure 1.17 and Table 1.2.

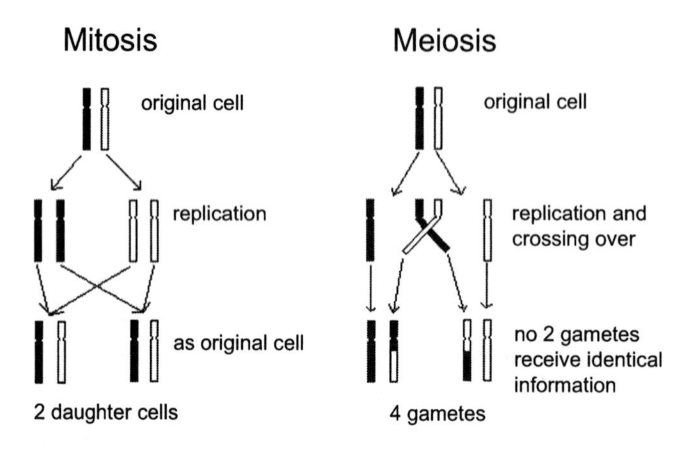

Figure 1.17 *Mitosis and meiosis*

Table 1.2 *The differences between mitosis and meiosis*

Mitosis	Meiosis
one division	two divisions
results in 2 daughter cells	results in 4 daughter cells/gametes
genetically identical	genetically different
same chromosome number	chromosome number halved
occurs in all body cells	occurs only in germ cells
occurs throughout life	occurs only after sexual maturity
used for growth and repair	used for production of gametes

ACTIVITIES 1.2, 1.3 AND 1.4

1.2. Explain why there is significant genetic variation as a result of meiosis but not of mitosis.

1.3. Describe the phases of the cell cycle.

1.4. Explain the reason why germ cells have to undergo meiotic division.

GENETIC INFORMATION

Chromosomes are made up of long chains of DNA (Deoxyribonucleic Acid) and protein molecules. It is the DNA within the chromosomes that holds all genetic information. The total length of the DNA within each cell is over 2m (6 feet) and, in order to fit within the cell's nucleus, it has to exist in a tightly packaged form. This is achieved by the DNA being coiled around protein structures called **histones** (see Figure 1.18). The DNA wraps around eight histones to form a structure called a **nucleosome**. Thousands of nucleosomes are formed, which gives the DNA molecule the appearance of a string of beads. Further coiling of these nucleosome beads results in a shortened structure called a **chromatin fibre**. It is these tightly packaged chromatin fibres that make up chromosomes.

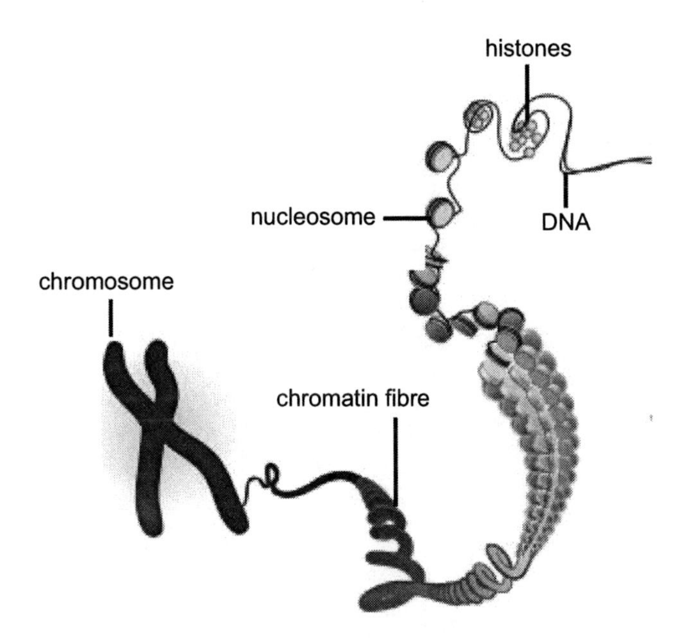

Figure 1.18 *Histones, nucleosomes and chromatin fibre*

The DNA within the chromosomes contains coded instructions for the production of protein. The coded area for the production of a specific protein is called a **gene**.

The structure of DNA

The structure of DNA was discovered through X-ray diffraction back in 1953 by the Nobel Prize-winning scientists James Watson and Francis Crick. DNA is composed of bases, sugars and phosphates that combine together to form a double helix. The double helix shape looks like a twisted ladder. The 'sides' of the ladder are made of phosphates and sugars, while the 'rungs' of the ladder are made of bases. Only four different types of bases exist within the DNA:

- Adenine (A);
- Guanine (G);
- Cytosine (C);
- Thymine (T).

DNA bases pair up with each other to form the 'ladder rungs' (see Figure 1.19). Adenine always pairs with Thymine, and Guanine always pairs with Cytosine. Only these two types of base pairing exist in DNA. The order of the base 'rungs' along the DNA ladder varies but the base pairings are always complementary.

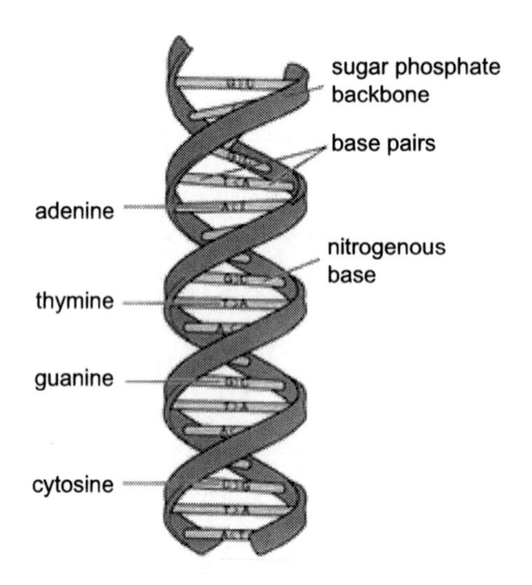

Figure 1.19 *DNA bases*

The sequences of bases on one DNA strand can be deduced from the sequence on the opposite strand, because base pairing is always complementary. Each strand independently carries the information required to form a double helix. Therefore, to describe a DNA sequence, only the sequence of the bases in one strand is needed, for example ATTGCAAT, as the other strand is always complementary, i.e. TAACGTTA. Human DNA consists of about 3 billion bases, of which over 99 per cent of the sequence is identical in all people. These bases, within the DNA, form the code for the production of proteins.

PROTEIN

All the functions of the cell depend on protein. Protein maintains cell structure, acts as both intracellular and extracellular messengers, binds and transports molecules and acts as enzymes.

Some proteins exist in every cell, such as the enzymes involved in glucose metabolism. Other proteins are highly specialised and are only found in specialised cells, such as the protein myosin, found only in muscle cells, or the protein insulin that is only produced in pancreatic islet cells.

What are proteins?

Proteins are made up of long chains of **amino acids**. There are only 20 different types of amino acids but, by varying the order and amount of amino acids in the chain, thousands of different proteins can be produced.

Links within the chain of amino acids are called **peptide bonds**, while the chain itself is known as a **polypeptide**. A protein can contain one or more polypeptides. Both the structure and function of the protein depend on the sequence of the amino acids making up the polypeptide chains.

In order to function, cells need information to produce proteins and the ability to pass this information on to new cells during cell division. This important information is provided by the DNA.

How are proteins made?

Proteins are not made in the cell nucleus but by the ribosomes in the cell's cytoplasm. The coded information in the DNA has to be transferred out of the nucleus. This is done by the use of ribonucleic acid (RNA).

Step 1: Copying the code

Segments of the DNA within the chromosomes separate at specific points and the DNA code is copied. This copy is called the **messenger RNA** (mRNA). During this process Guanine pairs with Cytosine and Adenine pairs with **Uracil**. RNA does not have Thymine but this is replaced with Uracil. Once a copy has been made, the DNA reattaches and the mRNA makes its way out of the nucleus into the cytoplasm (see Figure 1.20).

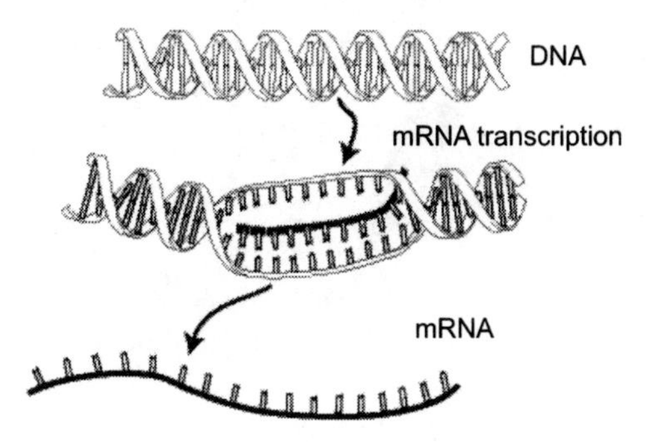

Figure 1.20 *Copying the code (transcription)*

Step 2: Reading the code

Once out in the cytoplasm, the mRNA attaches itself to a ribosome. Also present in the cytoplasm are amino acids, which are attached to a different type of RNA called **transfer RNA** (tRNA). The tRNA is only a short molecule of three bases that is attached to a corresponding amino acid. If the messenger RNA, which is attached to the ribosome, has three codes that correspond to the code on the transfer RNA, then the amino acid will be released by the transfer RNA. The released amino acid will then join other amino acids through the same process to form a protein molecule (see Figure 1.21).

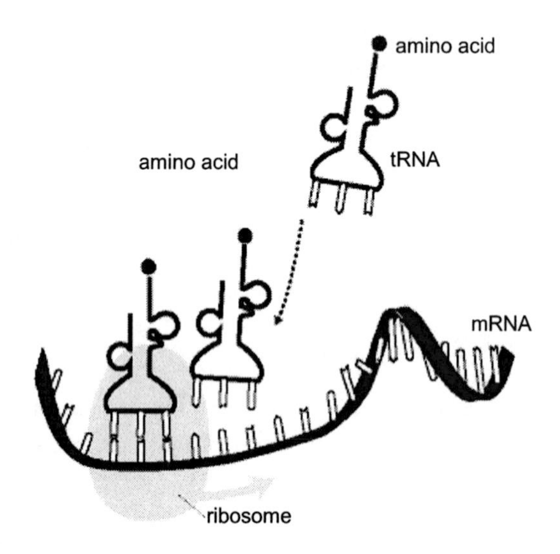

Figure 1.21 *Reading the code (translation)*

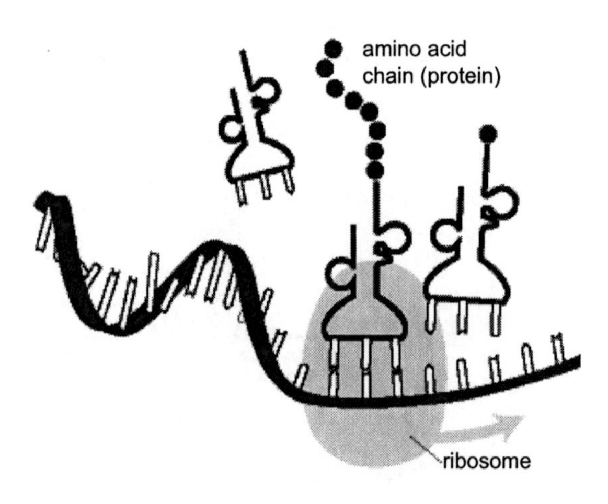

Figure 1.22 *Making the protein*

Step 3: Making the protein

When a peptide bond has been formed between amino acids they detach from the transfer RNA. The protein will now be constructed (see Figure 1.22).

Table 1.3 *Summary of the main processes*

Structure	Process	Function
DNA	None	Carries the genetic code
Messenger RNA (mRNA)	Transcription	Copies the code for a single protein from the DNA. Carries the copied code to the ribosomes
Ribosome	Translation	Reads the mRNA code and assembles the correct amino acid sequence
Transfer RNA (tRNA)	None	Brings individual amino acids from the cell cytoplasm to the ribosomes

The RNA code is written in a trinary code. Three bases code for one amino acid; this is known as a **codon.** There are four bases in RNA (Adenine, Guanine, Cytosine and Uracil) so a total of 64 possible combinations of codons can be achieved. As there are only 20 different types of amino acids, some amino acids can be coded for by more than one codon. This is referred to as **degeneracy** in the genetic code.

Some codons do not code for any amino acids but act as a start or stop signal. AUG (Adenine, Uracil, Guanine) has been recognised as a start codon and UAG, UGA and UAA act as stop codons.

The RNA base code for all amino acids has been deciphered since the 1960s and is known as the universal genetic code (see Figure 1.23).

The universal genetic code is based on the codons from the RNA where Uracil has replaced the Thymine base in the DNA. There is only one type of DNA whereas there are three different types of RNA (Table 1.4).

- **mRNA**: messenger RNA is a direct copy of DNA that codes for specific amino acids.
- **tRNA**: transfer RNA carries amino acids from the cytoplasm to the ribosomes.
- **rRNA**: ribosomal RNA facilitates interaction between mRNA and tRNA.

Second Letter

FIRST LETTER	U	C	A	G	THIRD LETTER
U	UUU, UUC phenyl-alanine UUA, UUG leucine	UCU, UCC, UCA, UCG serine	UAU, UAC tyrosine UAA stop UAG stop	UGU, UGC cysteine UGA stop UGG tryptophan	U C A G
C	CUU, CUC, CUA, CUG leucine	CCU, CCC, CCA, CCG proline	CAU, CAC histidine CAA, CAG glutamine	CGU, CGC, CGA, CGG arginine	U C A G
A	AUU, AUC, AUA isoleucine AUG methionine	ACU, ACC, ACA, ACG threonine	AAU, AAC asparagine AAA, AAG lysine	AGU, AGC serine AGA, AGG arginine	U C A G
G	GUU, GUC, GUA, GUG valine	GCU, GCC, GCA, GCG alanine	GAU, GAC aspartic acid GAA, GAG glutamic acid	GGU, GGC, GGA, GGG glycine	U C A G

Figure 1.23 *The universal genetic code*

Table 1.4 *The main differences between DNA and RNA*

DNA	RNA
double stranded	single stranded
deoxyribose sugar	ribose sugar
includes Thymine	includes Uracil
exists in one form	exists in different forms

ACTIVITIES 1.5 AND 1.6

1.5. The following table shows the sequence of bases on part of an mRNA molecule:

Base sequence on mRNA	CCU CAA AGU GGU GUU CGA
Base sequence on DNA	

a. Complete the table to show the DNA base sequence.

b. By using the universal genetic code table in this chapter, identify which amino acids are coded for.

1.6. A particular strand of mRNA is 60 bases long. How many amino acids would this strand code for?

MITOCHONDRIAL DNA

Not all the DNA in the human cell is contained within the chromosomes in the cell nucleus. Mitochondria, in the cell's cytoplasm, have their own DNA (the mitochondrial genome). This very small amount of DNA in the mitochondria is only inherited from the mother. Mitochondrial DNA is not inherited directly from the father as the mitochondria are placed in the tail of the sperm, which does not penetrate the ovum. In exceptional circumstances, when the tail of the sperm does manage to enter the ovum, the mitochondria are destroyed in the very early stages of embryo development.

The DNA within the mitochondria encodes for proteins that are essential for mitochondrial structure and function. This is a very small genome and most of the mitochondrial proteins are coded for by the nuclear genome.

THE CLASSIFICATION OF GENETIC MATERIAL

For any cellular structure to be classified as genetic, it must display four characteristics.

1. Replication.
2. Storage of information.
3. Expression of the stored information.
4. Variation.

1. Replication: this is achieved through the cell cycle when chromosomes are replicated in order to produce new cells.

2. Storage of information: chromosomes store all the information needed for the production of proteins. The genetic material within cells does not necessarily express all the stored information in every cell, only what is appropriate for that individual cell. For example, eye colour is not expressed in every cell, only in the cells which make up the iris of the eyes and the protein actin is only expressed in muscle cells and not in any other type of cell.

3. Expression of the stored information: expression is a complex process. Information flow requires DNA, RNA and cellular proteins (see Figure 1.24).

4. Variation: Genetic variation includes rearrangements between and within chromosomes as well as 'crossing over' during meiosis. This gives rise to trait variations between individuals and populations.

Chromosomes are the body's genetic material as they possess all four characteristics.

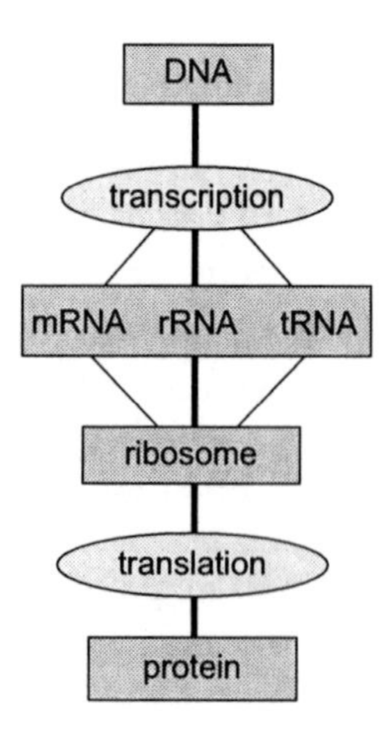

Figure 1.24 *Expression of stored information*

ACTIVITIES 1.7 AND 1.8

1.7. Explain and contrast a chromosome and a gene.

1.8. Identify the role of the following cellular components in the storage, expression and transmission of genetic information:

- chromatin;
- nucleus;
- ribosome;
- mitochondrion;
- centromere.

SUMMARY

- Cells are made up of organelles and chromosomes.
- Chromosomes are composed of DNA that encodes for proteins.
- There are 23 pairs of chromosomes in the nucleus of every cell (46 in total). There are 44 autosomes (22 pairs) and 2 sex chromosomes (X and Y).
- 23 individual chromosomes are inherited from each parent.
- Cells replicate to produce identical cells by mitosis. To halve the chromosomal number in germ cells, the cells replicate by meiosis.
- There are four different bases included in the DNA structure. A sequence of three bases (a codon) code for one amino acid. Amino acids link together to form protein.

- RNA is needed to copy and carry the genetic code out of the nucleus and to assemble the amino acid chain within the cytoplasm.
- Proteins are assembled following transcription and translation of the genetic code.
- Mitochondria have their own genome, although most mitochondrial proteins are coded for by the nuclear genome.

FURTHER READING

There are many good physiology texts that have a whole chapter dedicated to the biology of the cell.

Marieb, E. and Hoehn, K. (2006) *Human anatomy and physiology.* Harlow: Pearson International

Martini, F.H. and Nath, J.L. (2008) *Fundamentals of anatomy and physiology.* Harlow: Pearson International

Stanfield, C.L. and Germann, W.J. (2007) *Principles of human physiology.* Harlow: Pearson International

There are also some more in-depth texts on cellular biology.

Cooper, G.M. and Hausman, R.E. (2009) *The cell: A molecular approach.* Basingstoke: Palgrave Macmillan

Alberts, B., Bray, D., Hopkin, K., Johnson, A., Lewis, J., Raff, M., Roberts, K. and Walter, P. (2009) *Essential cell biology.* Oxford: Garland Science

For interactive web pages on cellular activities, the following websites provide some good animations in cellular activities.

www.cellsalive.com

www.biology.arizona.edu/cell_bio/cell_bio.html

www.johnkyrk.com/index.html

02

INHERITANCE

LEARNING OUTCOMES

The following topics are covered in this chapter:

- The Mendelian principles of transmission:
  - unit inheritance: genes and alleles;
  - dominance: allelic relationships;
  - segregation – single gene inheritance patterns, Punnet squares;
  - independent assortment – inheriting two or more genes.
- Exceptions to the rules:
  - mitochondrial inheritance;
  - penetrance;
  - genomic imprinting;
  - sex-related effects;
  - mutations;
  - genetic linkage;
  - polygenic and multifactorial inheritance;
  - epistasis;
  - pleiotropy.

INTRODUCTION

The fact that biological traits can be inherited has long been established. The first significant discoveries regarding the mechanisms of inheritance resulted from the work of Gregor Mendel in the late nineteenth century.

Mendel studied the patterns of inheritance within pea plants while he was working as a monk. His work went largely unnoticed until after the start of the twentieth century. Scientists who were studying the function of chromosomes rediscovered Mendel's publications and realised that Mendel had discovered the way in which biological traits were inherited. Mendel became known as the Father of Genetics, and the branch of genetics involved with simple inheritance is known as Mendelian genetics.

Although Mendel's work involved plants, his findings are relevant to human genetics. From his work, he derived certain laws that have become the principles of transmission genetics. Mendel proposed four principles of inheritance: unit inheritance, dominance, segregation and independent assortment. It is these four principles that form the basis of inheritance today.

1. The Principle of Unit Inheritance

Biological traits are determined by **genes**. Genes are the basic units of heredity. Strands of DNA that encode for one protein form a gene. As chromosomes occur in pairs after fertilisation, genes can be found on both the paired chromosomes.The individual 'genes' on each chromosome are termed **alleles.**[1] An allele is a version of a gene that has a paired version of the same gene in the same location on the opposite chromosome (see Figure 2.1).

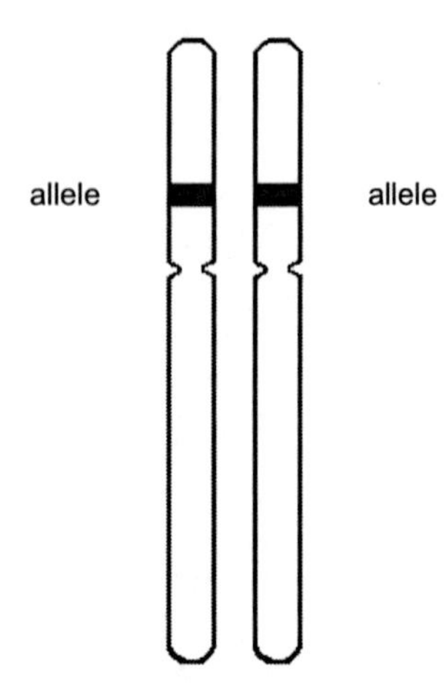

Figure 2.1 *A gene is usually made of two alleles, one on each of the paired chromosomes.*

2. The Principle of Dominance

Alleles can present as different versions of the same gene. If two alleles carried a different sequence of DNA, the effect of one allele might be masked by its partner allele.A **dominant** allele will be expressed regardless of any instructions carried by the other allele.

[1] Sometimes the words gene and allele can be used interchangeably.

In humans, the allele that codes for freckles is dominant over the allele for no freckles. Therefore, an individual who carries two different alleles for this gene – an allele for freckles and an allele for no freckles – will have freckles on their skin. This is because the freckles gene is dominant and will be expressed in that individual. An individual who has two different types of alleles for a single trait (like freckles) is said to be **heterozygous** for that trait.

An allele that is not expressed, due to the presence of a dominant partner allele, is termed **recessive**. Recessive alleles are only expressed when both alleles are in a recessive form. Individuals who have either two recessive alleles or two dominant alleles (i.e. two identical alleles) are said to be **homozygous** for that trait.

Whether an individual is said to be homozygous or heterozygous for a particular trait indicates whether they carry the same or different alleles within that gene. This can be described as the individual's **genotype**. A person's genotype is the genetic make-up for a particular trait. The term **phenotype** is used to describe the expression of the gene (or paired alleles) for the same trait (Table 2.1).

Table 2.1 *Genotypes and phenotypes for freckles*

Genotype		Classification	Phenotype
Allele 1	**Allele 2**		
freckles	freckles	homozygous	has freckles
freckles	no freckles	heterozygous	has freckles
no freckles	freckles	heterozygous	has freckles
no freckles	no freckles	homozygous	no freckles

Allelic relationships

Dominant alleles are phenotypically expressed in both heterozygotes and homozygotes. Recessive alleles are only expressed if the alleles are both in a recessive form (homozygous recessive).

Upper and lower case letters are used to represent dominant and recessive alleles. Upper case letters are used to represent a dominant allele and lower case for a recessive allele. If the letter 'F' was chosen to represent the gene for freckles, then 'F' would represent the dominant allele and 'f' would represent the recessive allele. An individual who is heterozygous for the freckles gene would be represented as an 'Ff' genotype. A homozygous dominant genotype would be 'FF', while a homozygous recessive would be 'ff'.

Any letter can be chosen to represent different allelic traits. However, it is good practice to choose a letter that has a different form in upper case compared with lower case. For example, A and a, B and b would be good to use but avoid C and c. This helps when drawing out inheritance patterns as different forms can be visually recognised as dominant or recessive, and it avoids errors due to poor handwriting.

Examples of Mendelian traits

Cleft chin

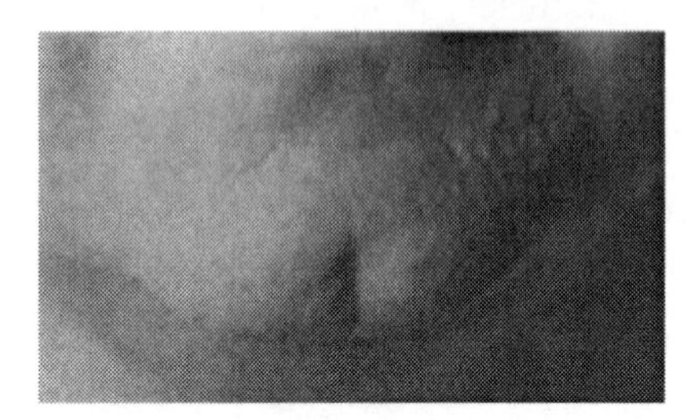

A cleft chin is due to a dominant allele (Figure 2.2). A person without a cleft chin has two recessive alleles for no cleft in their chin.

Figure 2.2 *A cleft chin*

Ear lobes

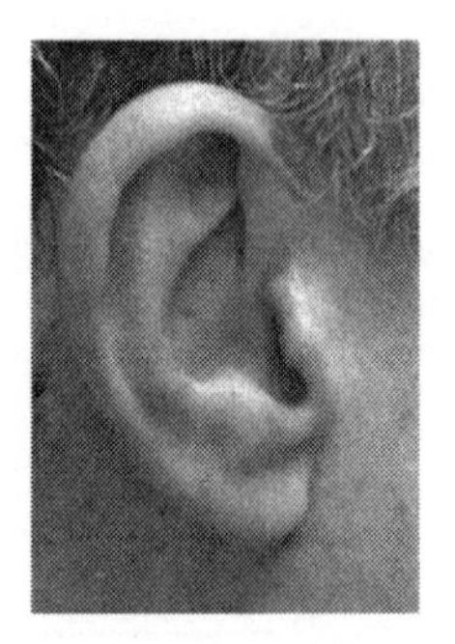

Free-hanging ear lobes, as in Figure 2.3, is a dominant trait.

Figure 2.3 *Free-hanging ear lobes*

Attached ear lobes, as in Figure 2.4, is a recessive trait.

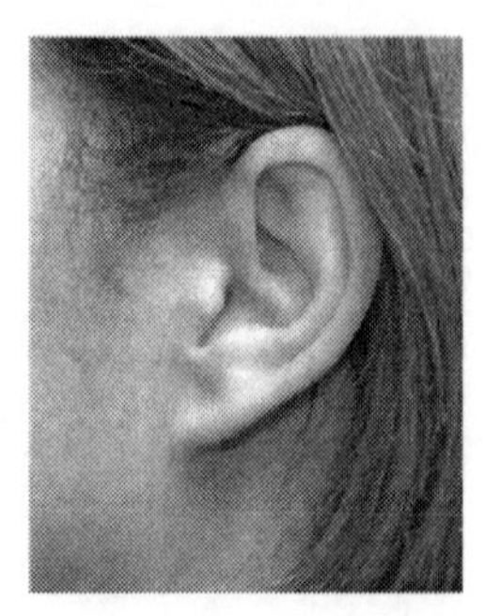

Figure 2.4 *Attached ear lobes*

Tongue rolling

The ability to form a U shape with the tongue is a dominant trait (Figure 2.5).

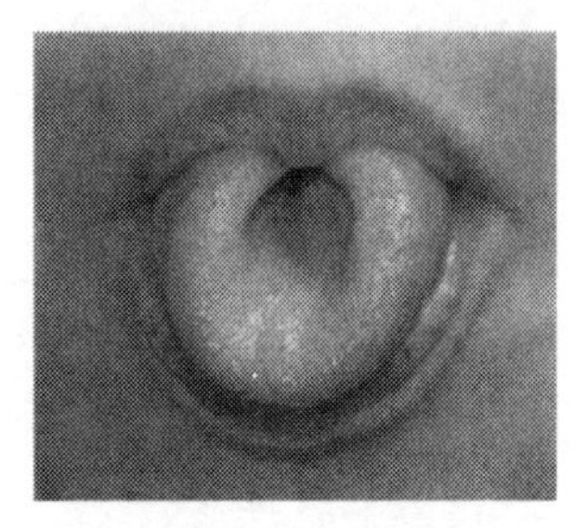

Figure 2.5 *Tongue rolling*

Widow's peak

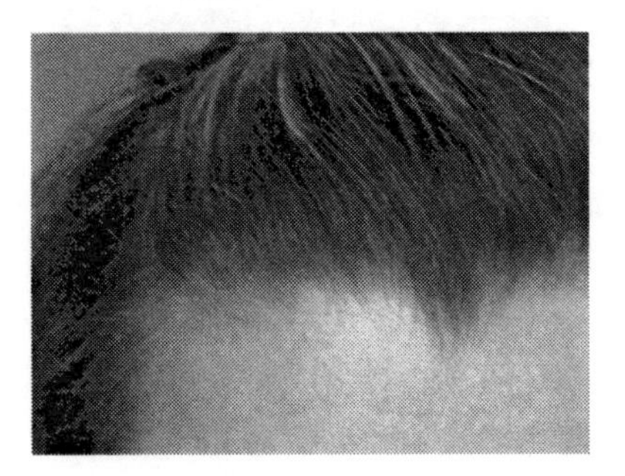

Figure 2.6 *Widow's peak*

When the hairline forms a V shape on the forehead, it forms a widow's peak (Figure 2.6). This is a dominant trait. Any individual who does not have a widow's peak is homozygous recessive for a straight hairline.

Dimples

Having dimples is due to a dominant gene (Figure 2.7). Individuals who do not have dimples when they smile have two recessive genes.

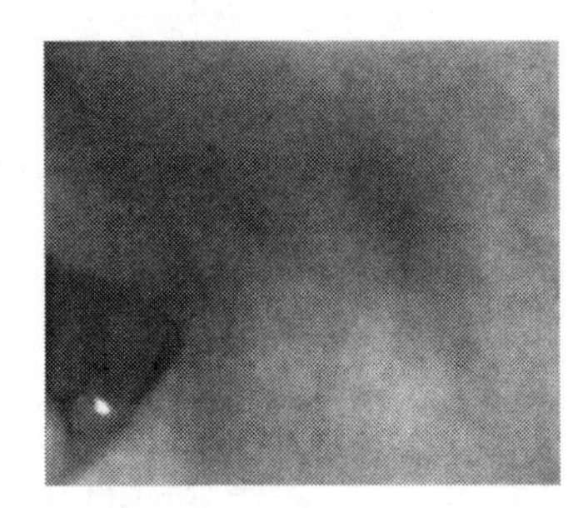

Figure 2.7 *A cheek with dimples*

Hitchhiker's thumb

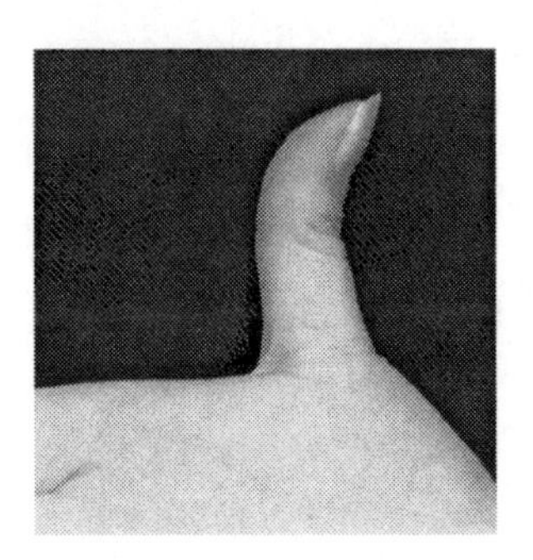

Figure 2.8 *Hitchhiker's thumb*

The ability to bend the thumb forward is due to a dominant gene (Figure 2.8). Individuals who have straight thumbs have two recessive alleles.

Carriers

A carrier refers to an individual who 'carries' a recessive allele for a particular trait but does not express that trait due to the presence of a dominant allele. Carriers are heterozygous, in that the recessive allele is present but is not expressed. An individual who carries a dominant and recessive allele for the freckles gene (heterozygous) has freckles but is also a carrier of the 'no-freckles' allele.

ACTIVITY 2.1

a. Which of the following would be a possible abbreviation for a genotype?

AB Cd Ee fg

b. Do the letters AA describe a heterozygous individual or a homozygous individual?

c. How many alleles for one trait are normally found in the genotype of an individual : 1, 2 or 3?

3. Principle of Segregation

During gamete formation alleles separate so that the gametes contain only one allele of each pair. Allele pairs are restored again after fertilisation.

All the nucleated cells in the body, except for the germ cells (sperm and ova), contain 46 chromosomes. These chromosomes consist of 22 paired autosomes and two sex chromosomes. Mendel's experiments were only on the traits carried by the autosomes of the plants and, therefore, the principles that he postulated apply to the 22 paired autosomes in humans (see Chapter 4 for sex-linked inheritance).

Somatic cells have the full set of paired chromosomes and are diploid (two copies of each chromosome). Germ cells have only half that amount (haploid) as none of the chromosomes are paired. The separation of the chromosomal pairs occurs during meiosis, leading to the formation of a haploid gamete. When fertilisation occurs between a sperm and an ovum to produce a zygote, the two sets of unpaired chromosomes unite to form a diploid zygote. Alleles combine in the offspring (see Figure 2.9).

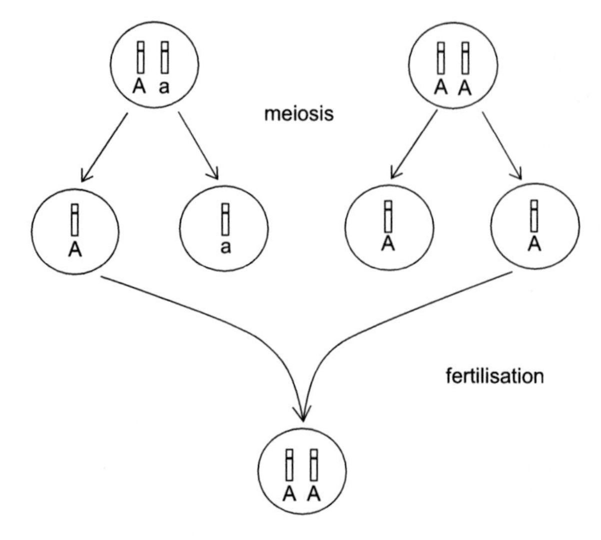

Figure 2.9 *Meiosis*

Working out the different allele combinations in the offspring is straightforward with single gene inheritance. The union of gametes that carry identical alleles will only produce a homozygous genotype (see Figure 2.10).

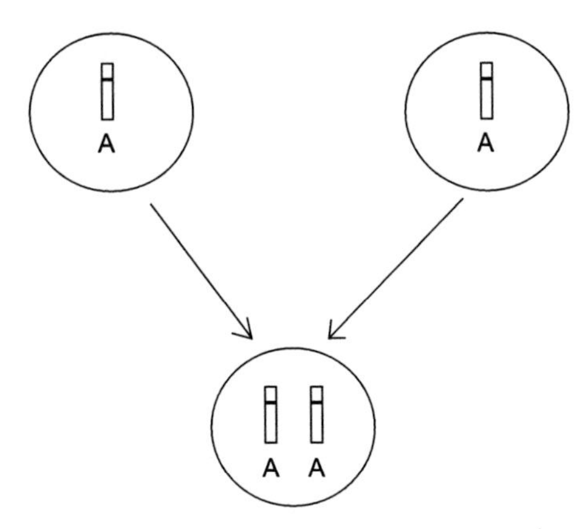

Figure 2.10 *Single gene inheritance*

For example, a mother who is homozygous recessive for straight hairline (not a widow's peak) and a father who is also homozygous recessive for this trait will only produce an offspring who is also homozygous recessive and will also have a straight hairline (see Figure 2.11).

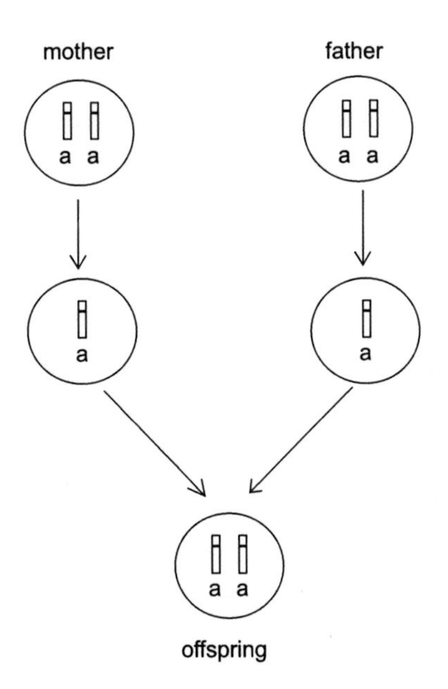

Figure 2.11 *A homozygous recessive genotype 'a' represents the recessive gene for straight hairline. The dominant gene 'A' for widow's peak is not present.*

The union of gametes carrying different alleles for the same gene will produce an offspring with a heterozygous genotype (see Figure 2.12).

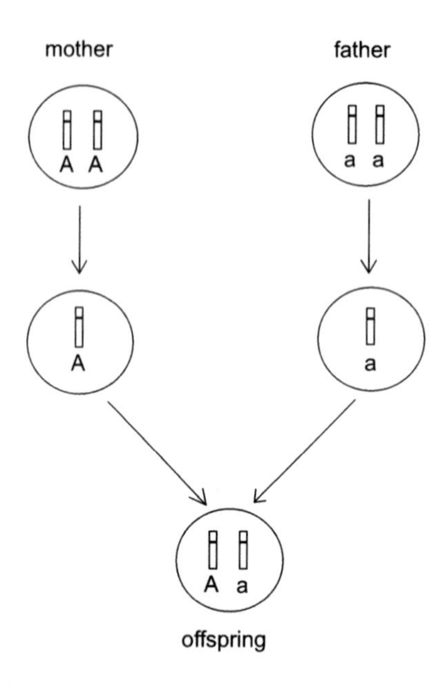

Figure 2.12 *A heterozygous genotype*

There are in fact six basic types of mating for single gene inheritance (Table 2.2). The examples that follow (Figures 2.13i–vi) are for ear lobe shapes, although these examples apply to all single gene recessive and dominant traits. Some individuals have 'free' ear lobes while others have elongated attachment of the lobe to the neck. Both 'free' ear lobes and 'attached' ear lobes are determined by different alleles of the same gene. The allele for free ear lobes is dominant over the allele for attached lobes. The letter chosen to represent the alleles in this instance is 'E' (E for ear lobes). 'E' represents the dominant 'free' lobe allele and 'e' represents the recessive 'attached' lobe.

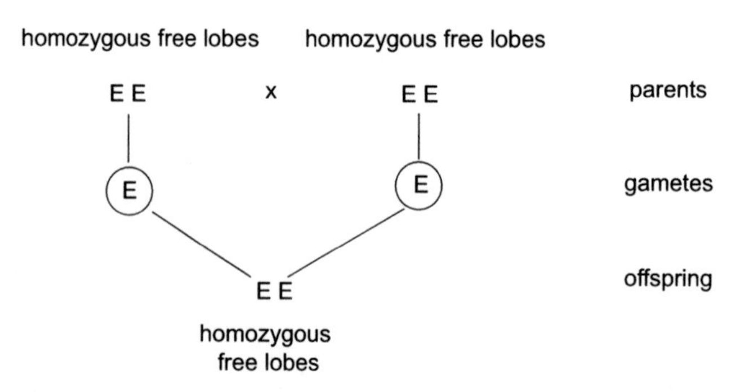

Figure 2.13i

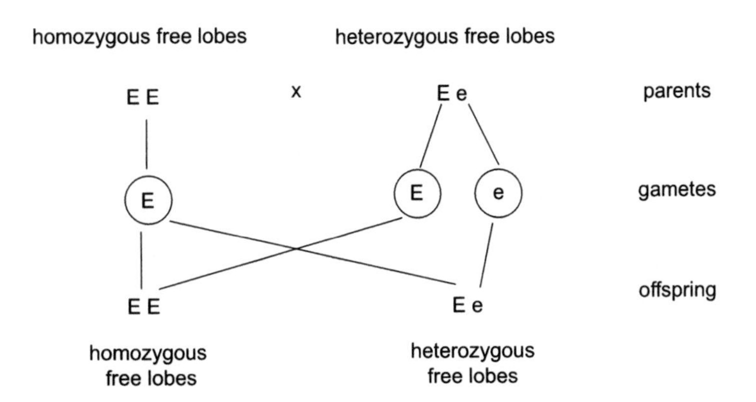

Figure 2.13ii

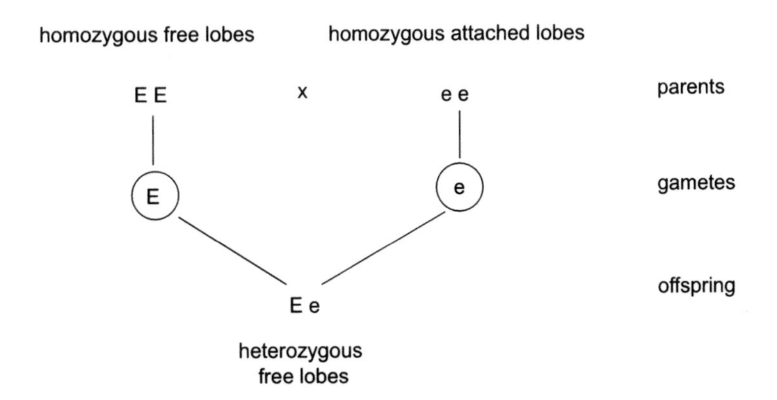

Figure 2.13iii

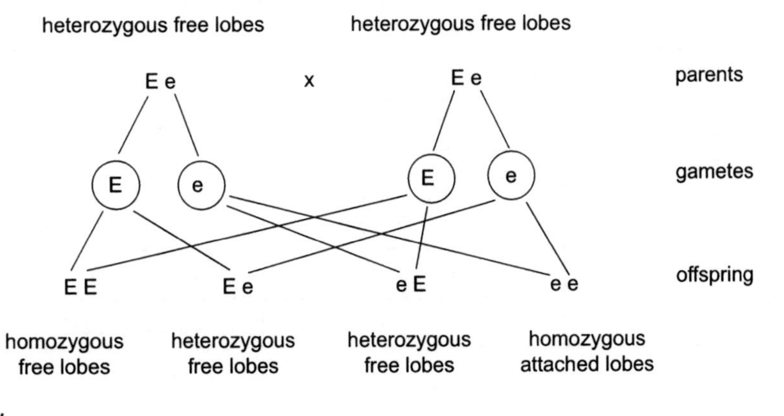

Figure 2.13iv

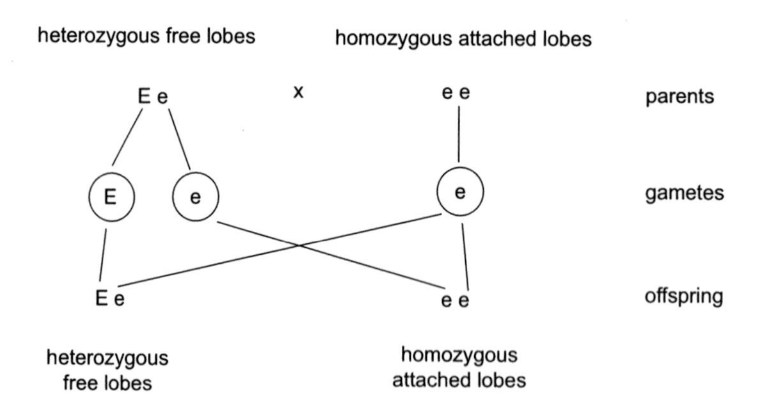

Figure 2.13v

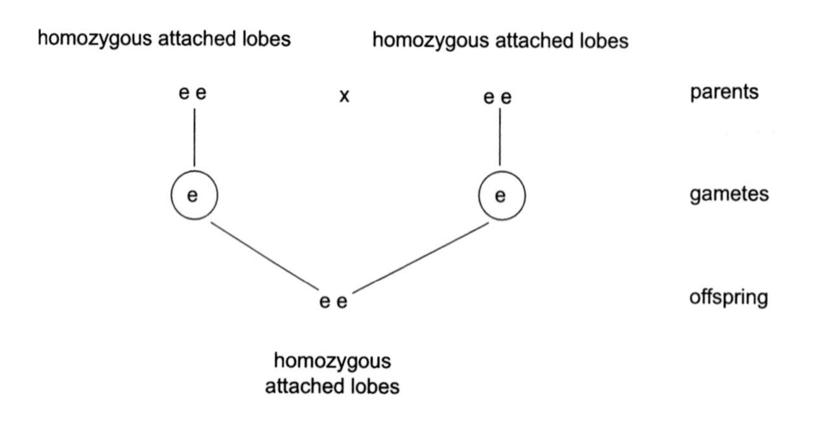

Figure 2.13vi

Table 2.2 *Summary of the six basic types of single gene inheritance*

Number	Parents	Genotypes	Phenotypes
i	EE x EE	100% EE	100% free lobes
ii	EE x Ee	50% EE, 50% Ee	100% free lobes
iii	EE x ee	100% Ee	100% free lobes
iv	Ee x Ee	25% EE, 50% Ee, 25% ee	75% free lobes, 25% attached lobes
v	Ee x ee	50% Ee, 50% ee	50% free lobes 50% attached lobes
vi	ee x ee	100% ee	100% attached lobes

Punnet square

An alternative method for working out the possible genotype of an offspring is the Punnet square.The Punnet square helps to visualise the segregation of alleles and the possible combinations within the offspring.

Drawing a Punnet square is quite straightforward once you realise that the parents' alleles segregate to form a gamete. The main framework consists of a grid composed of four perpendicular lines (see Figure 2.14i).

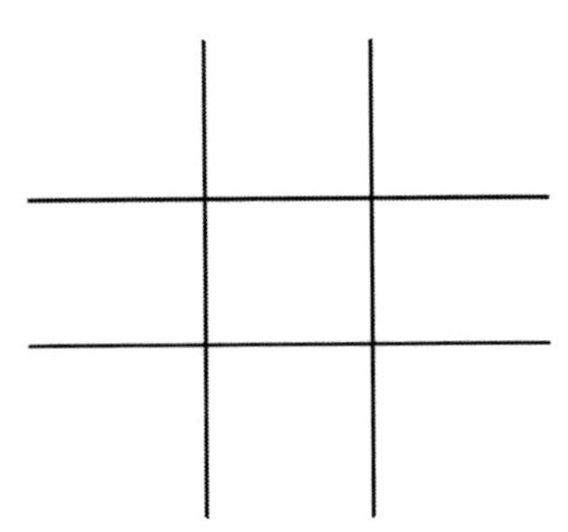

Figure 2.14i *Punnet square*

The genotype of one parent is then written across the top of the grid and the genotype of the other parent is written down the left-hand side of the grid. It makes no difference which parent's genotype is written at the top or the side (see Figure 2.14ii).

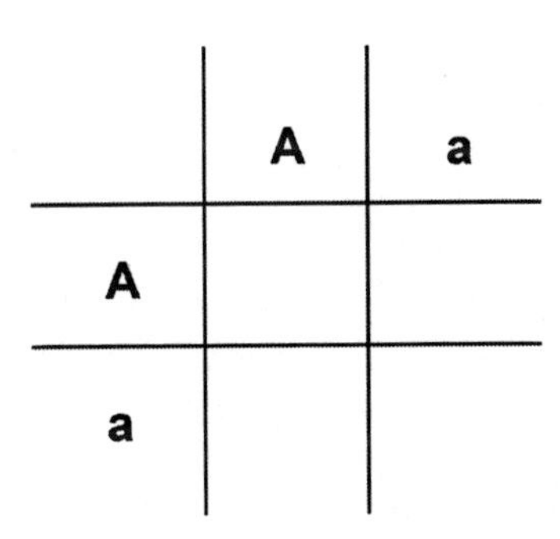

Figure 2.14ii *As alleles segregate to form a gamete, only one letter is inserted in each box at this stage*

By copying the column and row letters in each square, the possible combinations within the offspring can be worked out (see Figure 2.14iii).

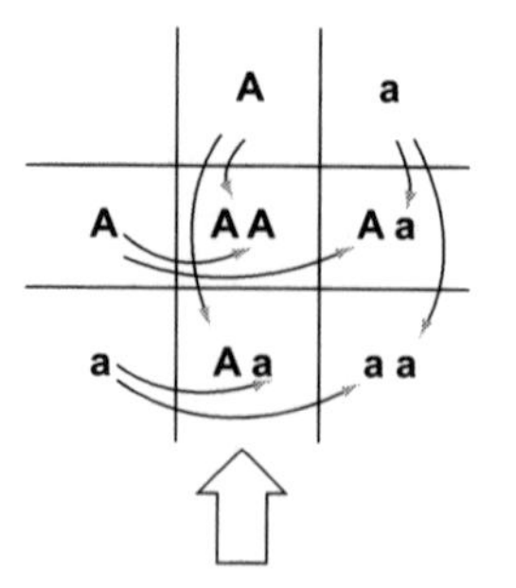

Figure 2.14iii *Punnet squares can be used to work out the possible genotype of offspring*

ACTIVITY 2.2

a. By drawing a Punnet square, what possible genotypes could the offspring of a Bb father and a bb mother have?

b. In the mating of two Bb individuals, what percentage of the offspring would have the same genotype as the parents? What percentage would have the same phenotype?

The maximum number of possibilities for a single gene inheritance is four (corresponding to the four squares in the Punnet square). However, these four possible outcomes can only contribute to a maximum of two phenotypes. In some situations there can only be one possible genotype and phenotype shared by all the offspring. For example, if one parent is homozygous dominant for a particular trait (GG) and the other parent is homozygous recessive (gg) the only possible outcome is for a heterozygous offspring (Gg).

The ability to form a U shape with the tongue is a dominant trait in humans. Consider the dominant allele being represented by the letter T and the recessive allele by the letter t. If two tongue rollers who were both heterozygous for this trait (Tt) had a child, what is the chance that the child would also be a tongue roller?

To work out this problem, a Punnet square needs to be drawn with the parents' genotypes inserted on the top and side of the square, and the possible offspring combinations inserted into the square (see Figure 2.15).

The results show:

- one homozygous dominant offspring (a tongue roller);
- two heterozygous offspring (tongue rollers);
- one homozygous recessive offspring (a non-tongue roller).

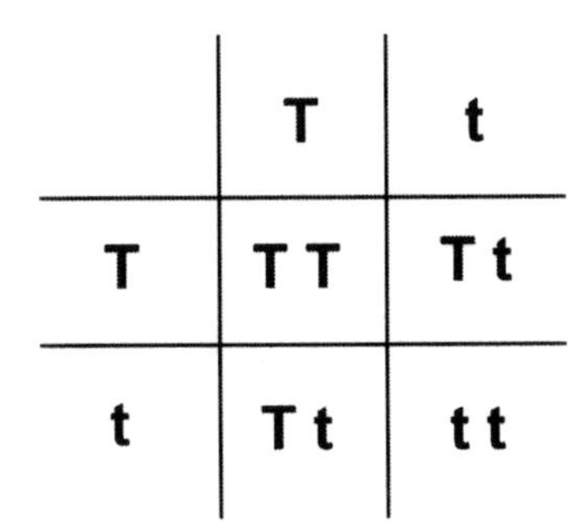

Figure 2.15 *Punnet square for tongue rollers*

As three out of the four outcomes are tongue rollers, the chance of having a tongue rolling child is 75 per cent.

ACTIVITY 2.3

Albinism is a condition that results in the lack of melanin pigmentation in skin. Individuals with this condition also lack pigmentation in both hair and the irises of the eyes. It is a recessive disorder and the condition only affects individuals if they have two recessive alleles for this condition (aa).

a. If two heterozygous individuals had a child together, what is the chance that one of their offspring will be albino? Work out your answer by drawing a Punnet square.

b. If a female carrier for the albino allele (she has normal skin colouring) has a child with an albino male, what are the possible genotypes and phenotypes for their offspring?

c. What are the chances that their offspring will also be albino like their father?

4. The Principle of Independent Assortment

Mendel's first three principles address traits that are inherited by single genes. Although there are quite a few genetic traits and conditions that are encoded for by a single gene, most are due to a number of genes that interact together. Since the sequencing of the human genome, scientists have discovered that single gene traits are relatively rare.

Mendel's fourth principle concerns the inheritance patterns of two different genes. The principle of independent assortment states that different genes control different phenotypic traits and the alleles reassort independently from each other. So even different genes within the same chromosome are independently assorted before the formation of a gamete. This occurs during the crossing-over of genetic material between chromosome pairs at meiosis (see Figure 2.16).

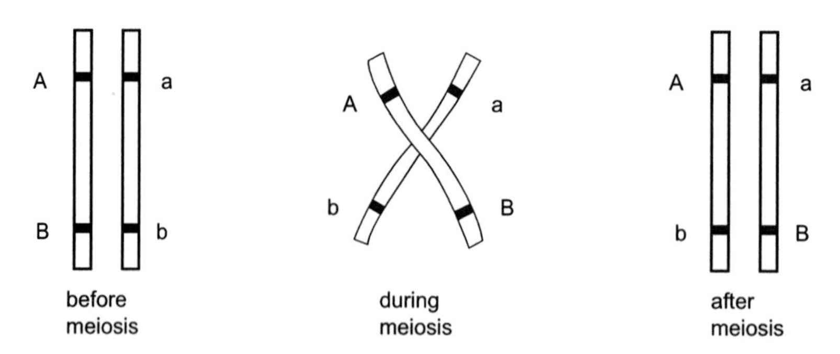

Figure 2.16 *Crossing over during meiosis*

So, the fourth principle considers genes transmitted on different chromosomes and that the transmission of one gene does not influence that of another gene. Independent assortment is explained through meiotic cell division. Chapter 5 looks at the inheritance of polygenic traits (multiple genes) in more detail.

ACTIVITY 2.4

The eldest son of two curly black-haired parents also has curly black hair. The middle son has straight black hair and the youngest son has curly blond hair. Which of the following Mendelian principles does this illustrate? Dominance, segregation or independent assortment?

EXCEPTIONS TO THE RULES

Mendelian genetics explains the rules of recessive and dominant inheritance. In the last 100 years the science of genetics has developed, giving us more understanding of how traits are inherited on a biological level. Although Mendelian principles still hold true today, there are a few exceptions to the rules.

1. Mitochondrial inheritance

In addition to the 46 chromosomes within the cell's nucleus, the mitochondria have their own genome. The mitochondrial DNA is not inherited in the same way as the nuclear DNA as it is only inherited from the mother and not from the father. Sperm never contribute mitochondria during the fertilisation of an ovum, so the mitochondrial genome within the ovum remains unchanged. This forms an exception to Mendel's law of segregation in that both parents do not contribute equally mitochondrial DNA to their offspring.

The mitochondrial DNA only forms a small part of the human genome. To date only 37 genes have been mapped to the mitochondrial DNA, most of which (24 genes) encode RNA molecules that are needed for protein synthesis within the cell's cytoplasm. The remaining 13 genes encode for proteins that are needed for cellular respiration. The mitochondria, although it has its own genome, is still reliant on genes from the nuclear genome to function adequately.

2. Penetrance

Penetrance relates to the expression of phenotypic features by a single gene. All Mendelian inheritance has a 100 per cent penetrance, but not all inheritance occurs in a Mendelian fashion. The degree of penetrance is measured in percentages. For example, achondroplasia (dwarfism), which is inherited in an autosomal dominant fashion, shows 100 per cent penetrance. This means that all individuals who carry this dominant allele will display the effects of achondroplasia. Other dominant autosomal genes are not always expressed and are said to have reduced penetrance. Ectrodactyly, a condition where the central parts of the hands and feet are not adequately formed, is an example of reduced penetrance. Not all individuals who carry this dominant gene will have deformities in the hands and feet. Degrees of penetrance are measured according to how many people display the phenotype of the gene in question. The BRCA 1 gene defect, which can cause breast cancer, is measured at 75 per cent penetrance, in that 75 per cent of individuals who have this genotype will develop breast cancer and 25 per cent will not.

3. Genomic imprinting

The imprinting of genes is a mechanism where the expression of a gene is governed by whether it was inherited from the mother or the father. Imprinted genes do not fit into the usual rules of inheritance as the contribution from one parent has been silenced. Both dominant and recessive genes can be imprinted. These genes are 'marked' with the sex of the parent that contributed it. There is no change to the actual DNA structure within these genes but a molecule of methyl is attached to the gene.

This process starts during gamete formation when certain genes are imprinted in either the developing sperm or the developing ovum. After fertilisation, the resulting offspring will have the same set of imprinted genes from both parents in all their somatic cells. However, the inherited imprinted genes will lose their methyl markers in the offspring's germ cells (sperm or ova). The inherited markers are removed in the germ cells and are 'reset'. This is done so that the new markers correspond to the offspring's own sex. A particular gene can therefore be turned on or off as it is passed through successive generations, from male to female to male.

The function of imprinted genes is not well understood. One possible reason for imprinting genes might be due to their role in embryonic development. Some genes lose their markers after birth, which suggests that imprinted genes may have an important role in regulating protein synthesis during pre-natal development.

Imprinted genes are important for normal development and health. If an imprinted allele is not silenced the cell receives two active copies of the allele and this results in over-expression of that gene. Similarly, if both alleles are imprinted the result is under-expression of that gene. This is the reason that parthenogenesis (virgin birth) is not possible in humans. An offspring needs both male and female genes so that the right proportion of genes is activated.

4. Sex-related effects

Some phenotypic traits are not inherited in a Mendelian fashion as they may be influenced by the sex of the individual. Some traits will only be expressed in one sex and not another, like, for example, beard growth. This is an example of a **sex-limited trait**, as beard growth is limited to males (even though both sexes inherit the gene).

Other traits can act as dominant in one sex and as recessive in the opposite sex. An example is male-patterned baldness in men. The allele for baldness acts as a dominant allele in males but as a recessive allele in females. This is known as a **sex-influenced trait**. (Females do not usually go bald, even with two recessive alleles for the baldness trait, due to the absence of necessary hormones.)

5. Mutations

A mutation is a permanent change in the sequence of chromosomal DNA. Mutated genes can be inherited in a Mendelian fashion. However, mutations can occur by chance and alter the genetic trait inherited by the offspring. For example, two parents of normal stature having a child with achondroplasia (dwarfism). As achondroplasia is due to a dominant gene, the affected offspring must have inherited or developed a change within the parental DNA because neither of the parents has achondroplasia.

Dynamic mutations are progressive changes within the DNA that occur from one generation to the next. This usually involves expansion of the DNA molecule that encodes for a particular gene. The genetic disorder resulting from this mutation might not appear for a few generations until the DNA within the gene has reached a particular length. Fragile X syndrome and Huntington's disease are just two examples of genetic conditions caused by dynamic mutations. Chapter 6 covers mutations in more detail.

6. Genetic linkage

Genetic linkage refers to different alleles that are positioned closely together on the same chromosome. Mendel's experiments were mainly on traits found on different chromosomes. When genes are located close together on the same chromosome, Mendel's prediction of independent assortment does not hold true.

Linkage refers to the transmission of genes on the same chromosome. The closer they lie, the less likely it is that they will separate during cross over in meiosis (see Figure 2.17). Individuals that have mixing of maternal and paternal alleles on a single chromosome have a parental or recombinant gene. These linked genes are inherited together and do not produce the Mendelian ratios of inheritance.

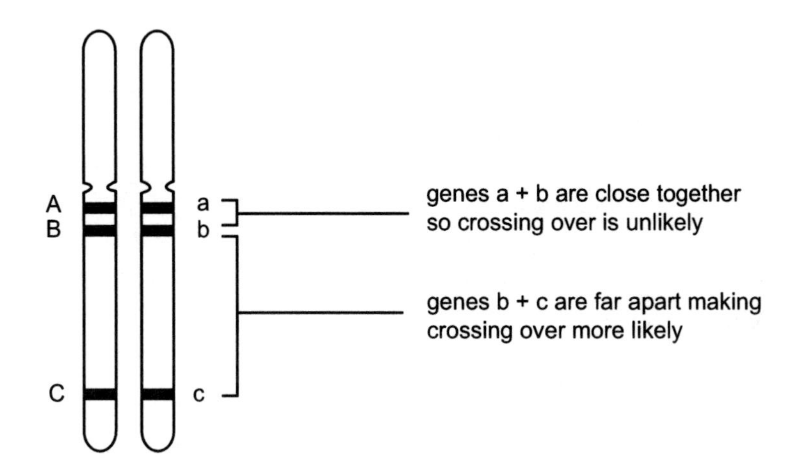

Figure 2.17 *Linkage*

7. Polygenic and multifactorial inheritance

Polygenic traits are controlled by the alleles of two or more genes without the influence of the environment. Many phenotypic traits are controlled in this way. However, both single gene and polygenic traits can also be multifactorial. Multifactorial traits are the result of both genetic and environmental influences. Most genes are actually multifactorial.

Human height is an example of a multifactorial trait. Genes are present that encode for height, but an adequate diet is needed to reach that height. Malnutrition in early childhood can have an effect not only on height but also on neurological development as well as other biological systems. Accidents at any stage of life can also alter an individual's phenotype. Multifactorial traits tend to follow Mendelian inheritance patterns, but the phenotypic results are difficult to predict due to the influence of the environment. Polygenic and multifactorial traits are discussed in greater detail in Chapter 4.

8. Epistasis

When one gene masks the effect of another unrelated gene it defies normal Mendelian inheritance patterns. Epistasis refers to the interaction of different genes, not between alleles of the same gene. A simplified example of epistasis in humans is that of the curly hair gene and the baldness gene. Although the curly hair gene is still expressed, it cannot have any effect on the phenotype of a bald individual.

A large number of proteins encoded by genes are involved in metabolic pathways. The absence of one step within the pathway will alter the outcome of the whole pathway. In epistasis, the remaining proteins needed to complete the pathway are present but are unable to interact in the pathway due to the 'missing step'.

The Bombay phenotype within the ABO blood grouping in humans is also an example of epistasis. ABO blood groups are due to the presence of A and/or B antigens on the surface of red blood cells. The A and B antigens are attached to the cell surface by proteins that are embedded in the cell membrane. The A and B antigen is encoded for by one gene and the protein attachments are encoded for by a different gene. If an individual does not have an effective gene to encode for the protein attachments, then the A and B antigens have no means of attaching to the cell surface. This individual would then display a blood group O phenotype, even though the genotype might be AB.

9. Pleiotropy

Pleiotropy is the expression of several different phenotypes by a single allele. Most single genes affect more than one observable trait. Pleiotropy occurs in genes which encode for a single protein that has more than one function within the body. Genetic disorders involving a pleiotropic gene are difficult to detect within families, as different members of the same family may display different symptoms. Marfan syndrome (see Chapter 7) is an example of a human condition arising from a pleiotropic gene, in that members of the same family can display different phenotypic traits arising from the same gene.

ACTIVITY 2.5

a. Benign epidermolysis bullosa is a condition that arises from an abnormal gene that encodes for collagen. The effect of this faulty gene results in the loss of skin and hair and in abnormal nails and teeth. Which of the following exceptions to Mendelian inheritance is this an example of: sex-related inheritance, genomic imprinting or pleiotropy?

b. One form of blindness is the autosomal dominant retinitis pigmentosa. The faulty gene that causes this type of blindness is incompletely penetrant. What do you understand by this statement?

Geneticists have identified thousands of genes that can lead to different traits, conditions and diseases.As more are discovered it is becoming clear that the 'exceptions to the rules' identified here are relatively common. However, a large proportion of traits are inherited in the Mendelian fashion.

SUMMARY

- Gregor Mendel, the father of genetics, outlined the four principles of inheritance.The principles of unit inheritance, dominance, segregation and independent assortment form the basis of Mendelian genetics.
- The principle of unit inheritance involves the transmission of hereditary units called genes. Genes are made up of two alleles, inherited from both parents.
- The principle of dominance involves the action of individual alleles within a gene, in that they are either dominant (will be expressed) or recessive. Recessive alleles are only expressed if both alleles are in a recessive form (or by the absence of a dominant allele).
- The principle of segregation refers to the separation of allelic pairs during meiosis. Allelic pairs are restored again at fertilisation.
- The principle of independent assortment concerns the inheritance patterns of two different genes.Alleles and genes resort independently from each other.
- An individual's genotype refers to their genetic make up and their phenotype refers to the outward appearance or the measurable effect of that gene.
- The term homozygous refers to two alleles that carry the same DNA in the same gene and heterozygous refers to different alleles within the same gene.
- There are exceptions to the Mendelian principles that include:
  - mitochondrial inheritance – maternal inheritance only;
  - penetrance – not all dominant genes are expressed;
  - genomic imprinting – some alleles may be silenced;
  - sex-related effects – may be sex-limited or sex-influenced;
  - mutations – DNA alterations can occur by chance as well as being inherited;
  - genetic linkage – alleles that are positioned closely together on the same chromosome have a higher chance of being inherited together;
  - polygenic and multifactorial traits – more common than single gene traits and may be influenced by the environment;
  - epistasis – unrelated genes can mask the effect of a gene;
  - pleiotropy – one single gene can result in the expression of different phenotypes, as one protein may have more than one function.

FURTHER READING

Cummings, M.R. (2008) *Human heredity: Principles and issues.* USA: Brooks Cole, International edition.

This is a well-written text, which has a good chapter on the transmission of genes from generation to generation (pages 44–69).

Griffiths, A.J.F., Gelbart, W.M., Lewontin, R.C. and Miller, J.H. (2002) *Modern genetic analysis: Integrating genes and genomes.* Palgrave:Hampshire, W.H. Freeman and Company.

Despite this text not being specifically about human genetics, it still has some very relevant material.

A good internet resource is Biology Online, which outlines Mendelian principles:
www.biology-online.org/2/1_meiosis.htm

Further information on the conditions mentioned in this chapter can be found on the Online Mendelian Inheritance in Man website:
www.ncbi.nlm.nih.gov/omim

Different genetic conditions are allocated a reference number on this site:

- Benign epidermolysis bullosa can be found as #226650.
- Autosomal dominant retinitis pigmentosa is referenced as #268000.
- Albinism is #203100.

03

AUTOSOMAL RECESSIVE AND DOMINANT INHERITANCE

LEARNING OUTCOMES

The following topics are covered in this chapter:

- autosomal recessive inheritance;
- autosomal dominant inheritance;
- variations in dominance;
- classification of gene action;
- co-dominance;
- multiple alleles;
- lethal alleles.

INTRODUCTION

In Chapter 2 the main principles of inheritance were explained. This chapter focuses on the inheritance of autosomal single gene disorders. Over 10,000 human diseases are due to single gene alterations and, although rare, they affect one per cent of the human population. Single gene disorders are also known as monogenic disorders. Genetic disorders are caused by abnormal genes. Alleles that become altered over time can be passed on to future generations. These altered alleles can result in the production of a non-functioning protein. An altered allele is a mutated allele.

The inheritance pattern of an altered gene depends on whether the gene is situated on an autosome (chromosomes 1 to 22) or on one of the sex chromosomes (XX in females, XY in males), and whether the alleles of that gene are either recessive or dominant. Genetic conditions arising from a single gene can be inherited in one of four ways:

1. autosomal recessive;
2. autosomal dominant;
3. X-linked recessive;
4. X-linked dominant.

Only the inheritance patterns of genes on the autosomal chromosomes will be explained in this chapter. X-linked inheritance is discussed in Chapter 4.

When the DNA coding within a gene becomes altered in any way, the resulting gene product may also be affected. The production of an altered or non-functioning gene can give rise to a genetic condition that affects health and development. These altered or mutated genes can be inherited in a recessive or dominant fashion.

AUTOSOMAL RECESSIVE INHERITANCE

Two copies of the altered allele must be present for an individual to be affected by a recessive disorder. That individual would be classified as homozygous recessive for that disorder. Heterozygous individuals who only possess one altered allele and a normally functioning allele will not display the effects of the altered allele in their phenotype but are classified as carriers of the altered allele. Carriers are not affected by the recessive allele but are able to pass that affected allele on to the next generation. Individuals need both alleles to be in the recessive form for the expression of the recessive phenotype.

Most individuals carry a small number of recessive alterations within their genes that cause no symptoms. Recessive diseases are single-gene disorders arising from two malfunctioning alleles (mutant alleles) and appear in homozygous individuals. Most affected individuals have two heterozygous parents who are unaffected because they have one altered and one normal allele and are carriers of the disorder.

Rules of autosomal recessive inheritance

- Both males and females are equally affected.
- Gene expression can 'skip' several generations as carriers do not express the gene.
- Affected children can be born to non-affected parents.
- If both parents are affected, all children will also be affected.
- Affected individuals with homozygous non-affected partners will usually have normal children.

Inheritance patterns

Affected individuals (homozygous recessive) are produced via one of three different types of mating:

1. Two heterozygous parents: **Aa x Aa** (both parents are carriers) (see Figure 3.1).

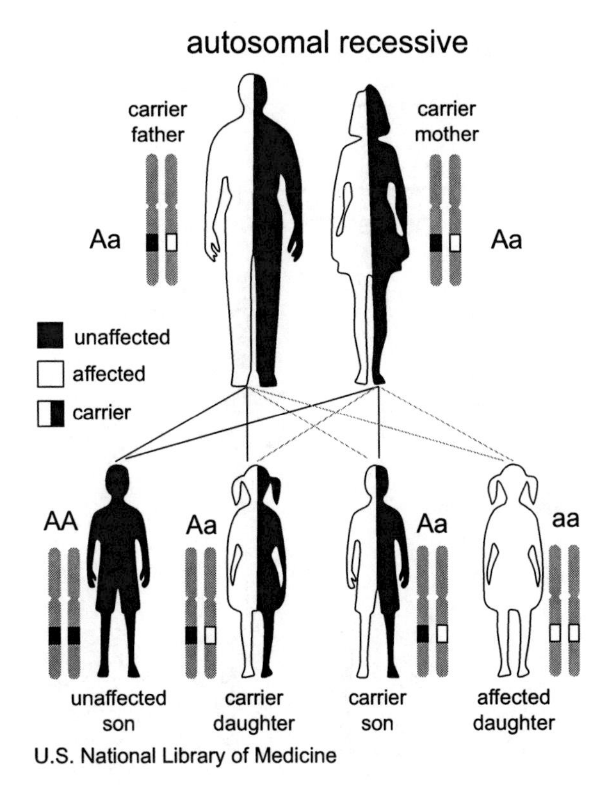

Figure 3.1 *Two heterozygous parents*
Key: **A** *= normal allele,* **a** *= affected recessive allele*

This is by far the most common type of mating that produces an affected offspring. The estimation of risk for an affected offspring from this type of mating is 25 per cent.

	A	a
A	AA	Aa
a	Aa	aa

Figure 3.2 *Estimation of risk from two heterozygous parents*

Offspring have a:

- 1 in 4 chance or 25 per cent risk of being an unaffected non-carrier (**AA**);
- 1 in 2 chance or 50 per cent risk of being a carrier (**Aa**);
- 1 in 4 chance or 25 per cent risk of being affected (**aa**).

2. Recessive Homozygote x Heterozygote **aa x Aa** (affected parent with a carrier parent) (see Figure 3.3).

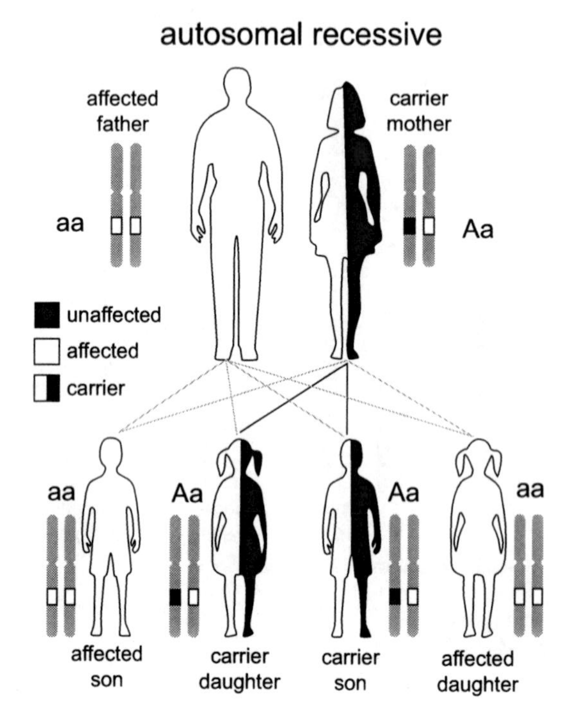

Figure 3.3 *Recessive homozygote x heterozygote parent*

The estimation of risk for an affected offspring from this type of mating is 50 per cent (Figure 3.4).

	A	**a**
a	**Aa**	**aa**
a	**Aa**	**aa**

Figure 3.4 *Estimation of risk from a recessive homozygote and heterozygote parent*

Offspring have a:

- 1 in 2 chance or 50 per cent risk of being a carrier;
- 1 in 2 chance or 50 per cent risk of being affected.

3. Two recessive homozygotes: **aa x aa** (both parents are affected) (see Figure 3.5).

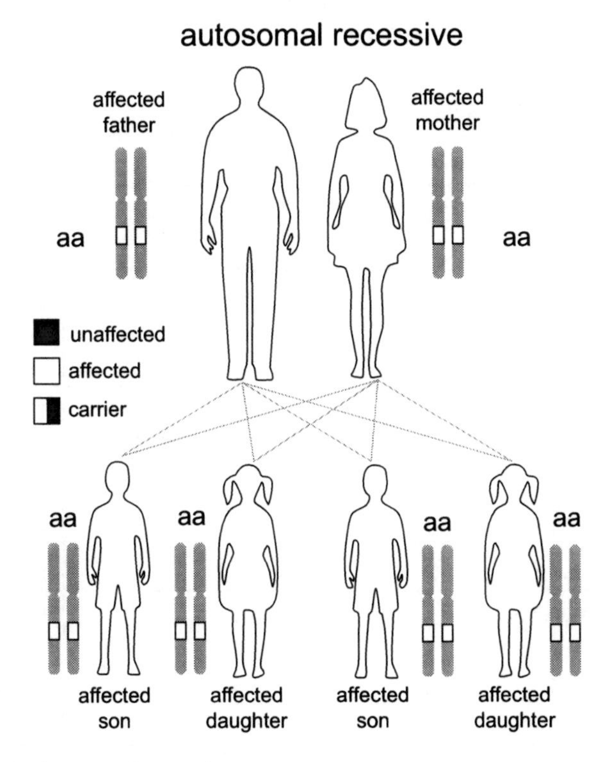

Figure 3.5 *Two recessive homozygote parents*

The estimation of risk for an affected offspring from this type of mating is 100 per cent (Figure 3.6).

	a	a
a	aa	aa
a	aa	aa

Figure 3.6 *Estimation of risk from two homozygote parents*

Offspring have a:

- 1 in 1 chance or 100 per cent risk of being affected.

Affected individuals who are homozygous recessive are usually the offspring of one of the above three matings.

There are thousands of autosomal monogenic recessive conditions. Table 3.1 contains a few examples of the most common conditions.

Table 3.1 *Common autosomal monogenic recessive conditions*

Condition	Chromosome	Gene	Effect
Adenosine Deaminase Deficiency	20q	ADA	Severe combined immunodeficiency
Batten Disease	16p	CLN3	Progressive disorder resulting in neuronal death within the brain
Congenital deafness	11p	USH1C	Deafness
Cystic Fibrosis	7q	CFTR	Defective chloride ion transport leading to thickened mucus production
Galactosaemia	9p	GALT	Developmental delay as a result of inefficient metabolism of galactose
Gaucher Disease	1q	GBA	Build-up of fatty deposits on liver, spleen, lungs and brain; anaemia and joint problems
Hereditary Haemochromatosis	6p	HFE	Iron overload due to too much iron being absorbed from the small intestine
Maple syrup urine disease	7q	DLD	Metabolic disorder leading to seizures, failure to thrive and developmental delay

(*Continued*)

Table 3.1 *(Continued)*

Oculocutaneous Albinism	11q	TYR	Lack of pigment in hair, skin and eyes
PKU	12q	PAH	Increased levels of phenylalanine leading to brain damage
Sickle Cell Anaemia	11p	HBB	Abnormal haemoglobin. Sickle-shaped red blood cells, which lead to the blocking of small blood vessels
Spinal Muscular Atrophy	5q 11 20	SMN1 IGHMBP2 VAPB	Progressive loss of function of motor neurones leading to atrophy of muscles
Tay-Sachs	15q	HEXA	Build-up of fatty deposits in the central nervous system, leading to death

Additional risks

Everyone carries several 'faulty' recessive genes that have no impact on their health. There are many different forms of faulty genes within a population but, because genes are inherited from parents and grandparents, family members will have more similarity within their genes and shared 'faulty' genes.

Consanguinity

The risk of developing an autosomal recessive genetic condition is increased in offspring of consanguineous relationships. The term **consanguinity** derives from the Latin prefix con-, meaning 'together', and the word *sanguis* which means 'blood'. It describes the marital relationship between two individuals who share a common ancestor. The most common form of consanguinity is the marriage between first cousins, which is encouraged in some cultures.

The children of unrelated parents are at low risk of inheriting two copies of the same faulty or altered allele. The risk of having a child with a birth defect is between 2 and 3 per cent, some of which will be due to a genetic condition. Children of parents who are blood relatives have an increased risk of having a genetic defect. The risk is doubled for parents who are cousins (5 to 6 per cent).The risk of inheriting the same faulty gene from both parents is increased the closer the relationship is between the parents (i.e. the more genes that they have in common) (see Table 3.2).

Table 3.2 *Relationships between blood relatives*

Relationship to each other	Brothers/sisters Parent/child	Uncles/aunts Nephews/nieces Grandparents Half-brothers Half-sisters	First cousins Half-uncles Half-aunts Half-nephews Half-nieces
Relationship type	First-degree relatives	Second degree	Third degree
Proportion of genes that they have in common	Half 50 per cent	Quarter 25 per cent	Eighth 12.5 per cent

The risk of having an affected child is much higher than 5 to 6 per cent in some families, because parents who are first cousins might also have grandparents who are themselves related.

ACTIVITY 3.1

a. A child who has a recessive genetic condition has two unaffected parents. If the child's genotype for this disorder is bb, what are the genotypes of the parents?

b. Why do recessive conditions appear to 'skip' generations?

AUTOSOMAL DOMINANT INHERITANCE

Autosomal dominant single gene disorders occur in individuals who have a single altered copy of the disease-associated allele. An alteration in only one of the alleles within a gene is enough to cause the disorder. The mutated disease-causing allele can be inherited from either parent.

Alleles encode for the production of a specific protein. When one allele is altered, in that the specific protein is no longer produced, the remaining functioning allele will still continue to encode for the specific protein. In autosomal dominant disorders, the amount of protein being encoded for by the functioning allele is not enough for the body to function normally.

In these cases the faulty allele causes a problem for the individual as it is dominant in its effect over the functioning normal allele.

In individuals who possess both alleles in an altered form (homozygous dominant), the disease symptoms are generally more severe. Dominant disease allele homozygotes are quite rare as many conditions appear lethal in the homozygous dominant form.

Rules of autosomal dominant inheritance

- Both males and females are equally affected, and can transmit to both sons and daughters.
- Most affected individuals will have an affected parent. The disease does not 'skip' generations.
- In affected families, where one parent is affected, the risk of transmitting the trait to the offspring is 50 per cent.
- If both parents are unaffected, none of the children will be affected.

Inheritance patterns

Affected individuals, who possess a dominant allele, are produced via one of three different types of mating.

1. Two homozygous dominant parents: **AAxAA** (both parents are affected) (see Figure 3.7).

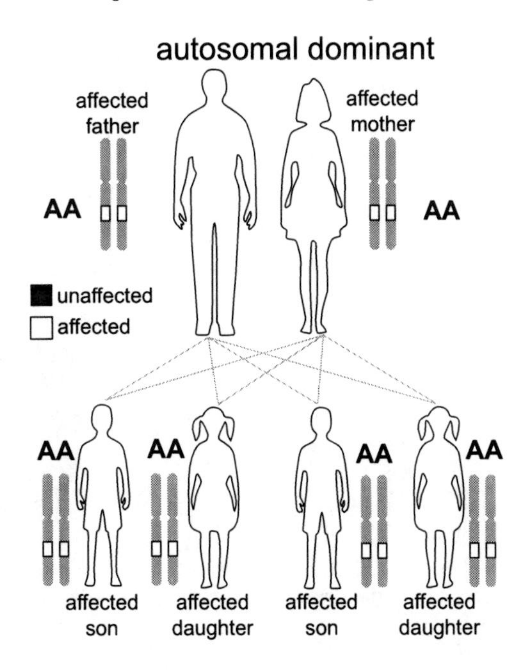

Figure 3.7 *Two homozygous dominant parents*
Key: **A** = dominant affected allele, **a** = recessive normal allele.

The estimation of risk of an affected offspring is 100 per cent (see Figure 3.8).

	A	A
A	AA	AA
A	AA	AA

Figure 3.8 *Estimation of risk from two homozygous dominant parents*

Offspring have a:

- 1 in 1 chance or 100 per cent risk of being affected.

2. Two heterozygous parents: **Aa x Aa** (both parents are affected) (see Figure 3.9).

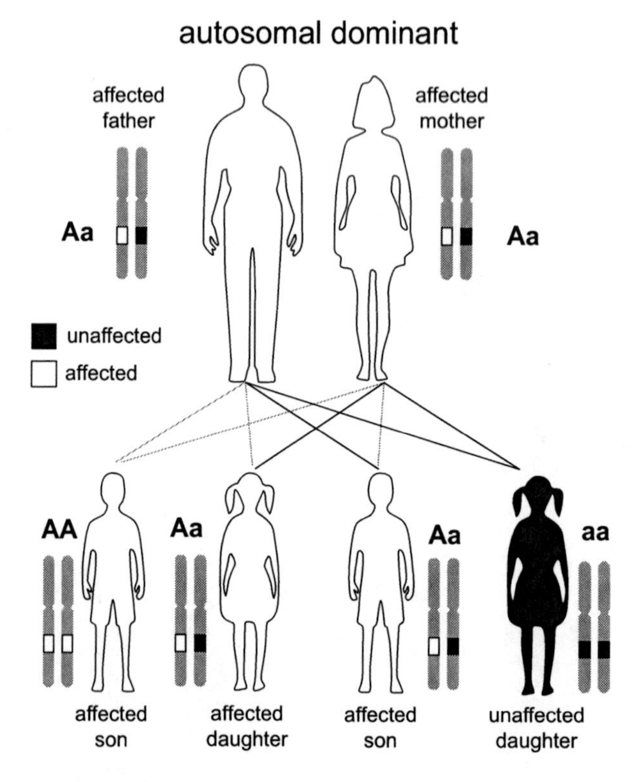

Figure 3.9 *Two heterozygous parents*

The estimated risk of having an affected child is 75 per cent (Figure 3.10).

	A	a
A	AA	Aa
a	Aa	aa

Figure 3.10 *Estimation of risk from two heterozygous parents*

Offspring have a:

- 3 in 4 chance or 75 per cent risk of being affected;
- 1 in 4 chance or 25 per cent risk of being unaffected.

3. Heterozygous x Homozygous recessive: **Aa x aa** (one affected parent and one unaffected parent) (see Figure 3.11).

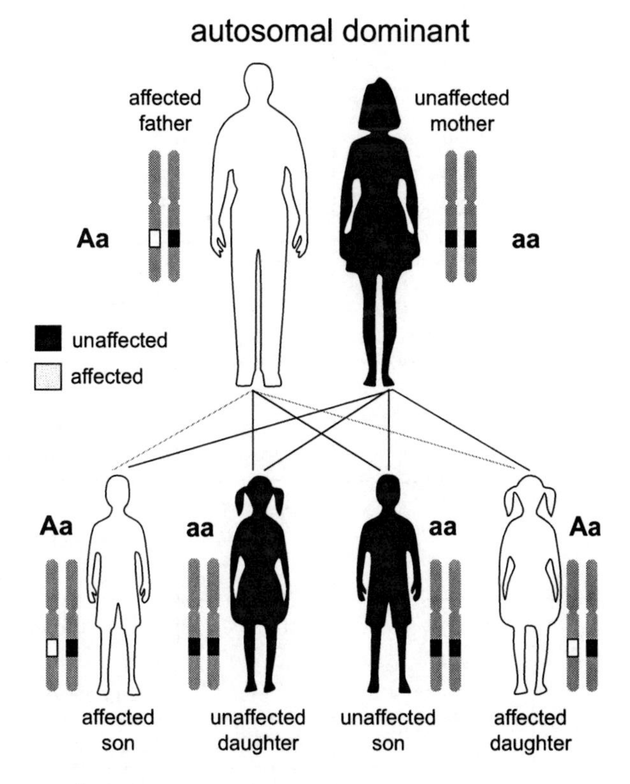

Figure 3.11 *Heterozygous x homozygous recessive parents*

Estimation of risk for this type of mating is 50 per cent (see Figure 3.12).

	A	a
a	Aa	aa
a	Aa	aa

Figure 3.12 *Estimation of risk from heterozygous recessive x homozygous recessive parents*

Offspring have a:

- 1 in 2 chance or 50 per cent risk of being affected;
- 1 in 2 chance or 50 per cent risk of being unaffected.

There are thousands of genetic conditions that are monogenic autosomal dominant. Table 3.3 gives some examples of the most common single gene dominant disorders.

Table 3.3 *Common monogenic autosomal dominant conditions*

Condition	Chromosome	Gene	Effects
Achondroplasia	4p	FGFR3	Dwarfism caused by severe shortening of the long bones of the limbs; lumbar lordosis and flattened bridge of the nose
Brachydactyly	9q	ROR2	Abnormally short phalanges (distal joints) of the fingers and toes
Huntington's disease	4p	HTT	Progressive brain disorder, involuntary movements and loss of cognitive ability
Hypercholesterolaemia	19p	LDLR	High blood cholesterol leading to increased risk of cardiovascular disease
Marfan Syndrome	15q	FBN1	Tall stature with elongated thin limbs and fingers; high risk of heart defects
Myotonic Dystrophy	19q	DMPK	Progressive muscle wasting

(*Continued*)

Table 3.3 *(Continued)*

Neurofibromatosis Type 1	17q	NF1	Growth of tumours along nerves in brain and skin; changes in skin colouration; increased risk of hypertension
Polycystic Kidney Disease Type 1	16p	PKD1	Fluid-filled cysts on enlarged kidneys and other organs, can lead to kidney failure
Polycystic Kidney Disease Type 2	4q	PKD2	Effects are the same as Type 1 but Type 2 has a later onset and symptoms are less severe
Porphyria Variegata	1q	PPOX	Inability to synthesise haem (essential for haemoglobin in red blood cells)

There are many more dominant traits than recessive traits recognised in humans. The reason for this is that a recessive trait can be 'hidden' by carriers whereas a dominant trait is always expressed. An individual with a dominant trait has a higher chance of having an affected child (a 50 per cent risk) compared with carriers of a recessive condition (a 25 per cent risk for two carriers).

ACTIVITY 3.2

a. Autosomal dominant conditions do **not** appear to 'skip' generations in the same way as autosomal recessive conditions. Explain the reasons for this.

b. What is the risk for two heterozygous dominant parents of having a child with the same condition?

c. Could a homozygous dominant affected individual and a homozygous recessive unaffected individual have an unaffected child?

For questions b) and c) you might need to draw a Punnet square (see page 31) to clarify your answers.

Variations in dominant inheritance

The way that dominant and recessive alleles behave is not always so straightforward. There are a few exceptions to the simplistic Mendelian inheritance patterns of dominance, even though the inheritance of these genes still follows Mendelian principles of inheritance.

1. New alterations

Most affected individuals with a dominant condition will have an affected parent. Some alterations in the chromosomal DNA can occur spontaneously either in the egg or sperm, or even early in embryonic development. Individuals may develop certain genetic conditions in this way. These individuals are affected by an altered allele, but their parents are not affected. The altered allele can be inherited by future generations. In some disorders the proportion of cases arising from new mutations is high. For example, 80 per cent of children born with achondroplasia do not have an affected parent but have developed the mutated allele either in early embryonic development or via a new arising mutated allele within the egg or sperm.

2. Late onset

Some autosomal dominant conditions are not expressed phenotypically until adulthood (e.g. Huntington's disease). This makes it difficult to predict risk when making reproductive choices.

3. Variable expressivity

The severity of symptoms of a dominant condition can vary between members of the same family, especially if the altered allele codes for a protein that is needed for different functions within the body. This makes it sometimes difficult to identify the condition and to track it through the generations of the family. Marfan syndrome has variable expressivity between members of the same family.

4. Incomplete penetrance

Usually a dominant allele will be phenotypically expressed. When an allele is always expressed it is said to be 100 per cent penetrant. There are some dominant conditions that do not follow this rule in that they have reduced penetrance. Retinoblastoma, an eye tumour, is an example of a genetic condition where the altered allele (allele RB on chromosome 13q) has variable penetrance. The susceptibility of developing the tumour is a dominant trait, but 20 per cent of individuals who have the altered allele do not develop the condition. The retinoblastoma gene therefore has an 80 per cent penetrance.

CLASSIFICATION OF GENE ACTION

Dominance usually occurs when a functioning allele is paired with a non-functioning allele. This usually arises from a mutation that alters the DNA structure within the allele, rendering it non-functional. An individual who has two altered alleles will generally display a distinctive phenotype as a result of the missing or altered protein produced by the altered alleles. It is not the lack of function that makes the allele recessive but the interaction of that allele with the alternative allele in the heterozygote. There are three main allelic interactions.

1. Haplosufficiency

This is when a single functional allele is able to encode for a sufficient amount of protein in order to produce a phenotype that is identical to that of the normal phenotype. If each allele encodes for 50 per cent of the amount of protein (100 per cent from both functioning alleles) and the normal phenotype can be achieved with only 50 per cent of the protein, then the functioning allele is considered dominant over the non-functioning allele. For example, the GALT gene on chromosome 9p that normally encodes for an enzyme needed for the breakdown of galactose shows haplosufficiency in the presence of one altered gene.

2. Haploinsufficiency

This is where a single functioning allele is unable to produce enough protein. Essential levels of protein must be over 50 per cent in cases of haploinsufficiency. The phenotype in haploinsufficiency resembles the homozygote for the non-functioning allele. This is rare in humans as deficiency usually results in a case of incomplete dominance.

3. Incomplete dominance

With a small number of alleles there is a lack of complete dominance. A heterozygous individual will have an intermediate phenotype compared with the two different homozygous individuals. The phenotype of the heterozygote becomes an intermediate or a 'blend' of the two different alleles. A simple example of incomplete dominance in humans can be seen with the gene for curly hair. An individual who has inherited a curly hair allele from one parent and a straight hair allele from the other parent will have wavy hair. In humans the 'blend' of the curly hair allele and the straight hair allele gives rise to wavy hair.

Most genes that display patterns of incomplete dominance have arisen from alleles in which a 'loss of function' has occurred. In a gene composed of one functioning allele and a non-functioning allele, only half the required amount of protein is encoded for by that gene. The genetic condition of familial hypercholesterolaemia demonstrates incomplete dominance in that individuals with one faulty or non-functioning allele will have raised blood cholesterol levels, while individuals who have two non-functioning alleles will have much higher cholesterol levels.

ACTIVITY 3.3

The straight hair allele (s) and the curly hair allele (c) show incomplete dominance in humans. Individuals with straight hair are homozygotes (ss), as are individuals who have curly hair (cc). Heterozygotes for this trait have wavy hair as they have one straight hair allele and one curly hair allele (sc). Note that the two different traits are represented by different letters.

a. Complete a Punnet square for a mating between a curly hair individual and a wavy hair individual.

b. What is the predicted offspring from this mating?

c. Is it possible for these individuals to have a straight hair child with each other?

d. Complete a Punnet square to determine the possible genotypes of the offspring of two wavy hair individuals.

e. Could two wavy hair parents have a child with straight hair?

f. Could the same wavy hair parents have a child with curly hair?

Whether an allele is classified as dominant or incomplete dominant depends on the individual's phenotype. However, the phenotype can be measured in different ways. Take, for example, the genetic condition of Tay–Sachs disease. Tay–Sachs disease is a degenerative condition that affects the nervous system. Affected individuals are born healthy but start to lose acquired skills at around the age of six months, gradually becoming blind, paralysed and unaware of their surroundings. It is a lethal condition with an average life expectancy of around five years. Affected individuals have two altered alleles in the HEXA gene on chromosome 15. A functioning HEXA gene is vital for development of the nervous system. Without the specific enzyme that this gene encodes for, fatty deposits build up in the brain, which then leads to neuronal damage. An affected individual has two non-functioning HEXA alleles. A heterozygote individual who has one functioning copy of the gene will be able to produce half the normal amount of the HEXA protein, which is enough to prevent damage from occurring. Heterozygotes are therefore carriers of Tay–Sachs disease. The healthy functioning copy of the HEXA allele is therefore classified as dominant to the non-functioning HEXA allele as the heterozygote individual displays no symptoms of the condition. However, if enzyme levels were measured, only half the usual amount of the HEXA enzyme protein would be discovered. In Tay–Sachs disease half the enzymatic levels are sufficient for health. At the biochemical level, the heterozygous individual displays incomplete dominance but complete dominance at the whole body level.

All the examples so far have demonstrated one allele being dominant or recessive over its partner allele. There are some conditions in which different versions of the same allele demonstrate equal dominance to each other. This is called **co-dominance.**

CO-DOMINANCE

Co-dominance is quite similar to incomplete dominance, in that neither of the two alleles is dominant or recessive to each other. However, there is no 'blending' in the offspring as both allelic products are expressed. Both parental traits are expressed in the offspring with co-dominant alleles. The biggest difference between incomplete dominance and co-dominance is that in co-dominance both alleles still encode for a functioning protein. The different proteins may have a slightly different function.

Most co-dominant alleles are thought to have arisen from a 'gain in function' mutation, where the alteration to the DNA structure within the allele has resulted in a different functioning protein being encoded for.

The MN blood group

An example of a co-dominant gene in humans is the gene that encodes for the MN blood group. The MN system is a type of blood grouping that is formed by the presence of specific antigens on the surface of the red blood cells. Two co-dominant alleles were originally identified for this blood group, termed M and N. The MN system is under the control of the MN gene located on chromosome 4. As both M and N alleles are co-dominant to each other there are three possible genotypes and phenotypes that can arise from the MN blood grouping system (see Table 3.4).

Table 3.4 *The MN blood grouping system*

Genotype	Phenotype
MM	MM blood group
NN	NN blood group
MN	MN blood group

There is distinct expression of both alleles in the MN blood group system, which is a characteristic of co-dominant inheritance.

ACTIVITY 3.4

In which of the following does the 'blending' of traits occur – incomplete dominance or co-dominance?

MULTIPLE ALLELES

The genetic inheritance patterns discussed so far have been limited to two alleles. The maximum number of alleles within a gene is two, one inherited from each parent. However, different forms of alterations in alleles can occur within populations, leading to numerous different forms of the same gene. When three or more forms of an allele exist for a single gene the term 'multiple alleles' is used. Note that multiple alleles can only exist in a population as an individual can only carry a maximum of two alleles within a gene.

Symbols used for multiple alleles

Multiple alleles also act in a dominant or recessive fashion, so capital and lower case letters are also used for multiple alleles. In addition, superscripts are used to aid identification within multiple alleles. The superscripts identify which form of the allele is present rather than its recessive or dominant action.

The ABO blood system

An example of multiple alleles in humans is the ABO blood system, in that there are three different types of alleles present in the population: A, B and O. An individual will only have two of these alleles within their individual genome, their blood group being dependent on the combination of the two alleles present.

In the ABO blood group system the possible blood groups that an individual might have are A, B, AB or O. This is determined by the expression of two out of the three possible alleles. As there are three different types of possible alleles, the gene for the ABO blood group is termed a 'tri-allelic' gene. This does not mean that the gene has three alleles, but has two alleles out of a possible three forms.

The alleles control the production of antigens on the surface of the red blood cells. Two of the alleles (A and B) are co-dominant to one another. The third allele (allele O) is recessive to the two co-dominant alleles as it does not encode for any antigens (acts as a 'loss of function' allele). The phenotype of an individual is determined by which antigens are present on the surface of their red blood cells (see Figure 3.13).

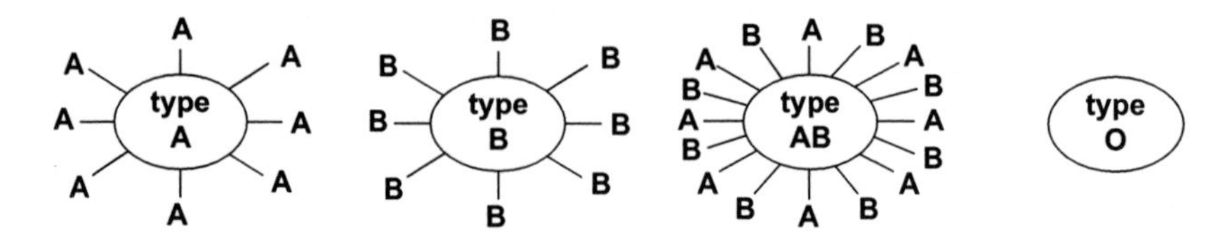

Figure 3.13 *Red blood cell antigens*

The symbols commonly used for the ABO blood system are: I for the representation of the allele (I for the dominant allele and i for the recessive allele). The letter I is used as it stands for isoagglutinogen, which is another term for antigen. The superscripts ^{A} and ^{B} are used to represent the encoded antigen (^{O} is not used as it does not code for any antigens).

- Allele I^A encodes for antigen A (blood group A).
- Allele I^B encodes for antigen B (blood group B).
- Allele i does not encode for any antigens (blood group O).

With three alleles, there is a higher number of possible combinations in a genotype (see Table 3.5).

Table 3.5 *Genotype combinations and phenotypes for blood group*

Genotype	**Phenotype**
$I^A I^A$	Group A
$I^A i$	Group A
$I^B I^B$	Group B
$I^B i$	Group B
$I^A I^B$	Group AB
$i\,i$	Group O

There are six different genotypes and four different phenotypes for the ABO blood system. The only homozygous recessive genotype is for blood group O.

Inheritance of blood group occurs following the Mendelian principles, taking into account the co-dominance of A and B and the dominance of A and B over O (see Figure 3.14).

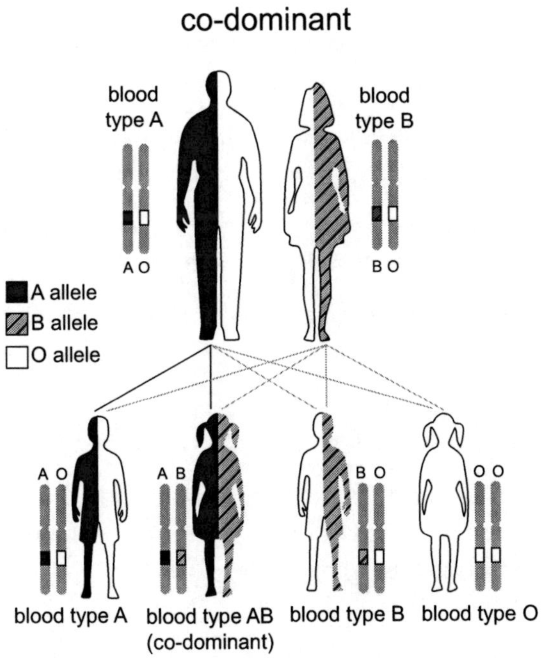

Figure 3.14 *The inheritance of blood group*

From a heterozygous mating between blood group A and blood group B individuals (AO x BO) the resulting offspring could be blood group A, B, AB or O (Figure 3.15).

	I^B	i
I^A	I^AI^B	I^Ai
i	I^Bi	i i

Figure 3.15 *Blood groups for the offspring of mating between individuals of groups A and B*

Offspring have a:

- 1 in 4 chance or 25 per cent risk of being group AB;
- 1 in 4 chance or 25 per cent risk of being group A;
- 1 in 4 chance or 25 per cent risk of being group B;
- 1 in 4 chance or 25 per cent risk of being group O.

The ABO blood group in humans is an example of co-dominance as well as being a multi-allelic trait.

ACTIVITY 3.5

The ABO blood group system in humans gives rise to four different types of blood; type A, type B, type AB and type O.

a. The A and B alleles are co-dominant with each other but dominant over the O allele. An individual with an AA or AO genotype will have type A blood.

i) List the possible genotypes that a person with type B blood could have.
ii) What is the genotype of an individual with type O blood?

b. Complete a Punnet square for a mating between a person with type AB blood and a person with type O blood.

i) From this mating is it possible to have a child with type AB blood?
ii) From the same mating is it possible to have a child with type O blood?

c. Complete a Punnet square for a mating between a heterozygous type A blood group individual and an individual who is heterozygous for type B blood. From this mating, is it possible to have a child with:

i) type AB blood?
ii) type O blood?

d. A mother who is blood group A has a blood group O child. The biological father of the child could be either of two men. One man has type B blood and the other has type AB. Who is the child's biological father?

e. Three babies have been 'mixed up' in the nursery on the maternity unit where they were born. Blood tests have revealed the blood types of the three babies and of their parents.

- Baby 1: type AB Parents 1: A and B
- Baby 2: type O Parents 2: AB and O
- Baby 3: type B Parents 3: AB and B

Work out which baby belongs to which set of parents.

LETHAL ALLELES

Any combination of alleles within a gene that results in the death of that individual is termed a lethal gene. Many of the proteins encoded for by different genes are essential for life. If one gene fails to 'work', the outcome might result in death. Death from a genetic disorder can occur at any stage of life. However, in terms of population genetics, a lethal gene results in the death of an individual before that individual has reached reproductive age. This prevents the gene from being passed on to future generations. The alleles of lethal genes can act in a dominant or recessive fashion.

Recessive lethal alleles

If the absence of a protein encoded by a gene results in death, it normally arises by a mutation leading to 'loss of function'. One allele that still encodes for the vital protein will often produce enough protein to compensate for the loss of function from the partner allele. If both alleles have mutated, resulting in total loss of function, then death occurs. Homozygous recessive individuals for a recessive lethal allele will not survive. The time of death varies according to when the normal gene product is essential for development. This could be at the embryonic stage, childhood or even adulthood. Gaucher Disease (perinatal form) and Tay-Sachs disease are both examples of genetic conditions with recessive lethal alleles.

Dominant lethal alleles

The presence of only one functioning allele in some genes will not be able to encode for a sufficient amount of vital protein for development. The non-functioning allele in this instance behaves in a dominant fashion as its loss of function will be displayed in the individual's phenotype and results in death. The genetic condition of Huntington's disease has a lethal dominant allele.

Sometimes a double dose of a dominant allele that causes a genetic disorder will result in an individual's death. For example, the altered allele that causes achondroplasia (dwarfism) behaves in a dominant fashion. A heterozygous individual will display the effects of the altered gene and will have the phenotypic characteristics of achondroplasia dwarfism. However, inheriting two altered achondroplasia alleles is rarely compatible with life. In this case the altered allele is dominant for the condition but acts as a recessive lethal allele.

ACTIVITY 3.6

Achondroplasia is a form of dwarfism. It is an autosomal dominant condition in that individuals only need the presence of one altered allele for this condition. Most individuals are heterozygous for this condition (Aa) as a 'double dose' of the altered allele (AA) is lethal.

Construct a Punnet square for a mating between two individuals who have achondroplasia.

a. Is it possible from this mating to have a child of normal height?

b. What are the risks for this couple in having a baby who will die from this condition?

c. What is the risk (in percentage form) of a father with achondroplasia and a normal height mother having a child with achondroplasia?

Many genetic conditions or diseases in humans are classified as either dominant or recessive. This, however, tends to be an oversimplified view of genetics. Recent estimates have put the number of protein-coding genes in the human genome at 25,000, of which approximately 1,800 are thought to be linked with single-gene disorders (monogenic disorders). Only a small proportion of these monogenic disorder genes have, as yet, been linked to specific diseases. Most of the common genetic disorders in humans arise from mutations in a number of different genes that interact closely together (polygenic disorders). Single-gene disorders (monogenic) are relatively rare compared with multiple-gene disorders (polygenic). Common genetic disorders tend to exhibit complex patterns of inheritance that involve interactions between a number of different genes, as well as having an environmental influence in the expression of the disorder.

SUMMARY

- Recessive genetic conditions are single-gene disorders arising from two malfunctioning alleles. Two copies of the altered alleles must be present in recessive conditions.
- Heterozygotes with one normal allele and a recessive non-functioning allele are carriers of the non-functioning allele. They are not affected by the altered allele.
- Recessive disorders can 'skip' generations as carriers are not affected by the recessive, disorder-causing, altered allele.
- The risk of being affected by a recessive condition is increased in offspring of consanguineous mating.
- Autosomal dominant disorders are caused by the presence of only one altered allele. Dominant disorders do not 'skip' generations as most individuals will also have an affected parent.
- Dominant alleles can arise from new mutations, have a late onset, variable expressivity and/or incomplete penetrance.
- Classification of whether an altered allele is dominant or recessive depends upon whether the partner allele of the non-functioning allele can produce enough gene product at a sufficient level for health and development. A lack of enough levels for health indicates that the normal allele is haploinsufficient. If enough protein is produced then the allele is haplosufficient.
- Incomplete dominance results in the 'blending' of traits.
- Co-dominance occurs when both alleles code for a different protein of which both are expressed.
- Multiple alleles can exist within a population of which an individual can have two varieties of that allele.
- Lethal alleles result in the death of the individual. They can act in a recessive (two copies needed for lethality) or dominant (one copy is lethal) fashion.

FURTHER READING

Bennett, R.L., Motulsky, A.G., Bittles, A., Hudgins, L., Uhrich, S., Doyle, D.L., Silvey, K., *et al.* (2002) 'Genetic counseling and screening of consanguineous couples and their offspring: Recommendations of the National Society of Genetic Counselors'. *Journal of Genetic Counseling*, 11(2), 97–119

This article sets out recommendations for genetic counsellors when working with blood-related couples of second cousin status or closer.

Cummings, M.R. (2008) *Human heredity: Principles and issues.* USA: Brooks Cole Publishing

This is a well-written text, which has a good chapter on the transmission of genes from generation to generation (pages 44–69)

Harper, P. (2004) *Practical genetic counselling*. London: Arnold

This is an excellent text, which provides good detail on the different modes of inheritance.

The National Genetics Education and Development Centre, which is part of the NHS, has a few web pages that outline the main principles of inheritance. These pages can be found at:
www.geneticseducation.nhs.uk/learning-genetics/patterns-of-inheritance.aspx

This site also provides a lot of other resources that you may find useful in practice.

Statistical information on the number of genes detected and identified within the human genome can be found through the statistical pages of the Online Mendelian Inheritance in Man website, which is hosted by the Johns Hopkins University:
www.ncbi.nlm.nih.gov/omim

Further information on the conditions mentioned in this chapter can also be found on this website.

04

SEX-LINKED INHERITANCE

LEARNING OUTCOMES

The following topics are covered in this chapter:

- the X chromosome;
- the Y chromosome;
- sexual development;
- inheritance of sex;
- sex ratios;
- sex-linked inheritance patterns;
- pseudoautosomal inheritance;
- X inactivation;
- sex-limited and sex-influenced inheritance.

INTRODUCTION

Within the human genome there are 46 chromosomes or 23 chromosomal pairs. Chromosomes are grouped in homologous pairs, that is, two chromosomes that share the same genes at the same locations in both chromosomes. The first 22 pairs of chromosomes (the autosomes) are truly homologous. The 23rd pair (the sex chromosomes) is not truly homologous in both sexes. These sex chromosomes are made up of the X chromosome and the Y chromosome and it is these chromosomes that determine an individual's gender. Both males and females have 22 pairs of autosomal chromosomes in every cell nucleus and a pair of sex chromosomes. Females have two X chromosomes and males have one X chromosome and one Y chromosome. The presence of a Y chromosome determines the gender as male (see Figure 4.1).

The X and the Y chromosomes do not look alike and do not share the same genes. Males, who have XY chromosomes, are termed the heterogametic sex (two different sex chromosomes) and females are termed the homogametic sex (two chromosomes that are

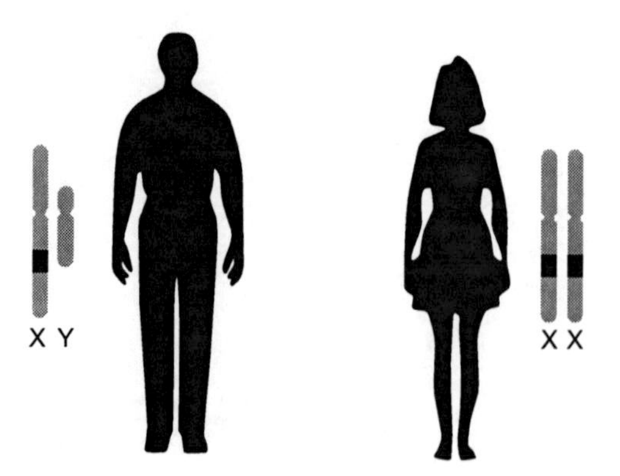

Figure 4.1 *The sex chromosomes*

alike). The presence of a Y chromosome in the genome indicates male and the absence of a Y chromosome indicates female.

THE X CHROMOSOME

The X chromosome is a much longer chromosome in comparison to the Y chromosome and it contains over 1,500 genes. Its structure is similar to those of the autosomes. In females, who have two X chromosomes, the chromosomes are homologous in that the alleles are paired and act in a dominant or recessive pattern. In males, who have an X and a Y chromosome, there is little in common, DNA-wise, between the two sex chromosomes, apart from two small regions at the tips of the chromosomes. Most of the genes carried on the X chromosome have very little to do with the inheritance of sex. All genes carried on the X chromosome are said to be X-linked and display different inheritance patterns between males and females (see Figure 4.2).

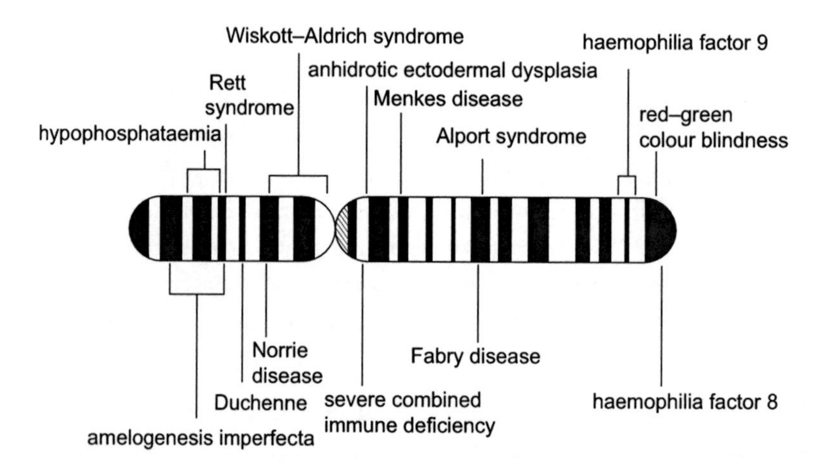

Figure 4.2 *The X chromosome and some disorder-causing genes*

THE Y CHROMOSOME

The Y chromosome is much smaller than the X chromosome, and contains only about 230 genes. This is far fewer than the X chromosome. The Y chromosome is unusually organised in that 95 per cent of the chromosome contains male-specific genes. The remaining 5 per cent of the genes are known as the pseudoautosomal genes and are situated on the tips of the Y chromosome. There is only a small number of pseudoautosomal genes (63 genes discovered to date) found in the pseudoautosomal regions of the Y chromosome (see Figure 4.3). These genes have counterparts on the X chromosome, and are able to cross over and exchange DNA between the two chromosomes during meiosis. Some of the genes found in the pseudoautosomal regions encode for proteins that are involved in bone growth, enamel formation on teeth, cell division, immunity, formation of hormones and fertility.

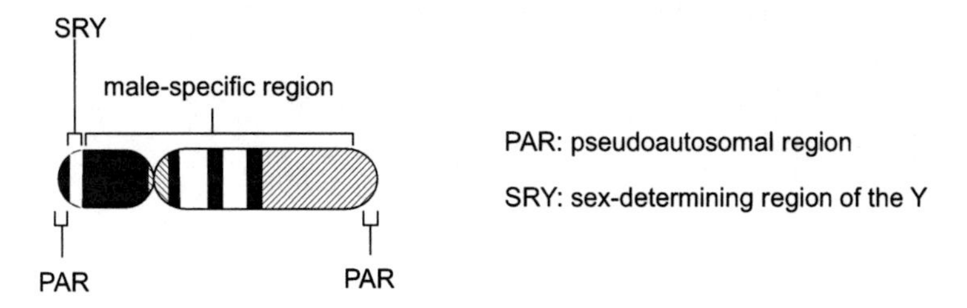

Figure 4.3 *The Y chromosome showing the different genetic regions*

The male-specific region makes up most of the Y chromosome. Within the male-specific region, adjacent to the pseudoautosomal region of the short arm of the chromosome, is a gene that is critical in male development. This gene is the sex-determining region Y (SRY). It is this gene that encodes for a protein that is involved in the development of testes in males. On very rare occasions some males are born with a Y chromosome and multiple X chromosomes (XXY, XXXY and even XXXXY - see Chapter 6 for multiple X conditions). However, as the Y chromosome carries the SRY gene for determining maleness, all individuals with these conditions are male. Genes carried on the Y chromosome are said to be Y-linked.

CASE STUDY 4.1

Olympic Testing

The presence or absence of the sex-determining region has been used to confirm gender in women's athletics.

SEXUAL DEVELOPMENT

The presence of a Y chromosome results in the development of a male. All embryos are genetically 'pre-programmed' to develop as females. At around the sixth week of pre-natal development the sex-determining region of the Y chromosome (SRY) becomes activated. Hormones then steer the development of the embryo along the male route. In the absence of the Y chromosome (and therefore the SRY) the embryo 'defaults' to produce female features.

The SRY in males begins a cascade of events that involves activating other genes, many of which are situated on the autosomes as well as on the sex chromosomes.

ACTIVITY 4.1

a. How do genes in the pseudoautosomal region of the Y chromosome differ from the genes in the male specific region?

b. What phenotype would the following individuals be?
- XX genotype.
- A person with a non-functioning SRY gene on the Y chromosome.
- XXY genotype.

THE INHERITANCE OF SEX

Females have two X chromosomes in every cell nucleus and males have one X chromosome and one Y chromosome. When germ cells are produced (sperm in males and ova in females), the chromosomal complement is halved through meiosis. Ova therefore only carry one X chromosome, but sperm may have one X chromosome or one Y chromosome (see Figure 4.4).

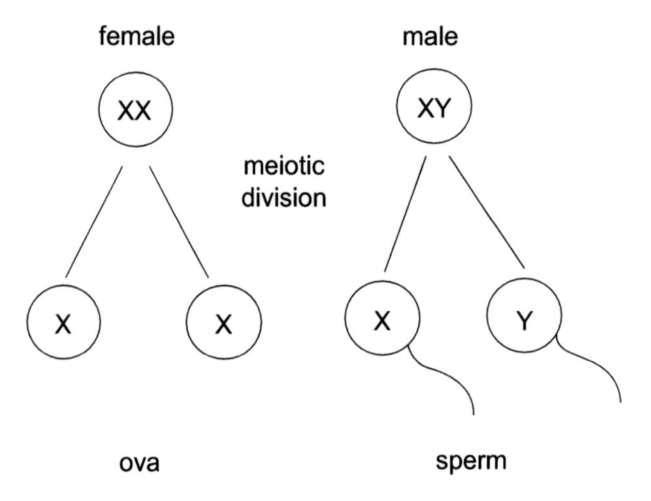

Figure 4.4 *Meiotic division*

Females are homozygous in respect to the sex chromosomes, as only one type of ovum can be produced, one that contains an X chromosome. Males are heterozygous in that two types of sperm can be produced; sperm containing an X chromosome or sperm containing a Y chromosome. An embryo that has inherited its father's X chromosome will develop into a female and an embryo that inherits a Y chromosome will develop into a male. A Punnet square can also be used for chromosomal inheritance and not just for the inheritance of single genes (Figure 4.5).

		male	
		X	Y
female	X	XX	XY
	X	XX	XY

Figure 4.5 *Punnet square for chromosomal inheritance*

The results of this Punnet square demonstrate that the mating results in a 1:1 ratio of sons and daughters. It is in fact the male that determines the sex of the offspring.

Sex ratios

By following Mendel's rules of segregation, there is a 50 per cent chance that an offspring will have inherited a Y chromosome from the father and be male, and a 50 per cent chance of inheriting an X chromosome from the father and becoming female. The ratio of males to females should, in theory, be 1:1.

However, this proportion of males to females within populations can be affected by other factors, such as survival at birth, as well as social and environmental factors. The sex ratio within populations is measured by calculating the number of males, then dividing this by the number of females within the same population. This division is then multiplied by 1,000. This calculation is done on individuals within certain age groups. For example, if a certain population consisted of 3,000 men and an equal number of women:

> 3,000 men divided by 3,000 women = 1
> This is then multiplied by 1,000:
> 1 x 1,000 = 1,000
> So a population that has an equal number of males and
> females has a sex ratio of 1,000.

The sex ratio at conception is termed the primary sex ratio, while the sex ratio at birth is termed the secondary sex ratio. It is not always possible to get an accurate calculation of the primary sex ratio due to lack of sex identification in some early aborted foetuses.

The tertiary sex ratio of a population can be divided into different age categories following birth. These ratios can alter with advancing age groups, reflecting that some medical conditions and environmental factors can affect one gender more than another. For example, the sex ratio tends to favour females towards the end of the human life span, with more females outliving men in older age.

ACTIVITY 4.2

a. If there were more males than females in a certain population, would the sex ratio be greater or less than 1,000?

b. An island population has 10,400 males and 12,500 females.

i) What is the sex ratio of this population?

ii) How many males are there for every 100 females in this population?

The World Health Organization has calculated a world sex ratio of 101.3 males to every 100 females. This data has been achieved by measuring and combining the sex ratios in individual countries. In the UK the ratio is 98 males to every 100 females. This measurement is for the total UK population, regardless of age. In fact there are 105 male births to every 100 female births in the UK but, due to the ageing population of the country (there are only 76 males to every 100 females in the over-65-year age group), more women than men survive into old age.

One of the ideas put forward as to why more boys are conceived in comparison with girls (see for example Case Study 4.2) is that the Y chromosome is so small; the sperm containing the Y chromosome weighs slightly less than the X-containing sperm. This could give the Y-containing sperm an advantage in that they could reach the ova more quickly.

CASE STUDY 4.2

Missing Females

Some societies display a significantly skewed sex ratio where there are many fewer female births than expected. There are various explanations for unequal sex ratios at birth including biological, environmental and cultural reasons.

In countries such as China and India, cultural practices have resulted in a greater number of males than females. In both countries there has been a cultural preference for baby boys over baby girls. Researchers such as Gupta (2005), Jha et al. (2006) and Zeng *et al.* (1992) have identified a large number of 'missing females'. This was first detected in China with the under-reporting of female births and female infanticide practices leading to the skewed sex ratio.

The same male bias is also present in areas within India, although much of this is now attributed to pre-natal diagnostic techniques leading to termination of XX foetuses. Although pre-natal sex determination has been illegal in India since 1994, the law is often ignored.

Jha *et al.* (2006) studied 1.1 million households in India and found that girls conceived into families that already had a daughter had a higher probability of being aborted or dying after birth. Girls without any older siblings had similar survival chances to boys. This indicated that cultural factors can explain some of the 'missing females'.

There are currently 100 million 'missing females' in India alone.

ACTIVITY 4.3

Missing males

What environmental or cultural factors could account for a skewed sex ratio resulting from fewer adult males than females?

SEX-LINKED INHERITANCE

Genes carried on the X chromosome are said to be X-linked and genes carried on the Y chromosome are Y-linked. Y-linked traits are only passed from father to son as only males have a Y chromosome. The X chromosome, however, is present in both sexes (two X chromosomes in the female genome, one in the male genome). In females, due to the presence of two X chromosomes, the X-linked traits are inherited in the same way as autosomal traits. Two copies of a recessive allele are needed for the expression of that trait. Dominant and recessive genes behave the same as autosomal dominant and recessive genes in females.

The male has only one copy of the X chromosome, which means that all the genes on that X chromosome will be expressed, whether they are dominant or recessive in nature. A man will inherit an X chromosome only from his mother; there is no direct male-to-male transmission of X-linked genetic material.

Fathers always pass the X chromosome to their daughters, never to their sons. Mothers pass their X chromosome to both daughters and sons.

X-linked inheritance

As the Y chromosome lacks most of the genes that are present on the X chromosome, the genes inherited on the X chromosome exhibit a unique pattern of inheritance. Recessive genes on the X chromosome are always expressed in males as there is no corresponding allele on the Y chromosome. Recessive expression in females only occurs if both alleles are in a recessive form. In females, the X chromosome acts in the same recessive/dominant pattern as the autosomes.

Some faulty alleles, which have arisen through mutations, can be found on the sex chromosomes. These faulty genes can give rise to certain medical conditions. The altered or faulty genes on the sex chromosomes can be inherited in a dominant or recessive fashion. There are two general rules for X-linked inheritance:

1. Sons of a carrier mother and a non-affected father have a 50 per cent chance of being affected (see Figure 4.6).

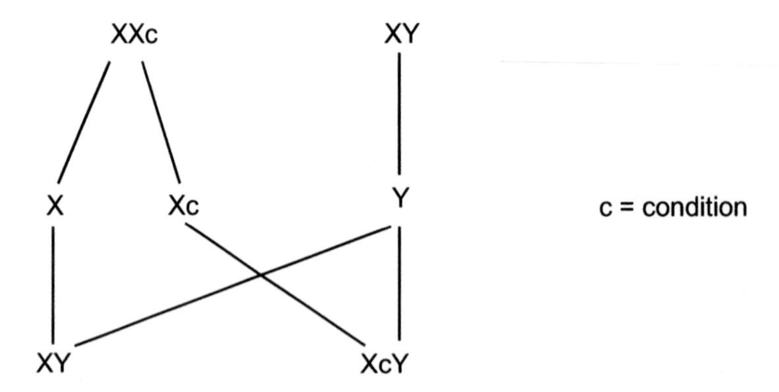

Figure 4.6 *X-linked inheritance 1*

2. Daughters of a carrier mother and a non-affected father have a 50 per cent chance of being carriers (see Figure 4.7).

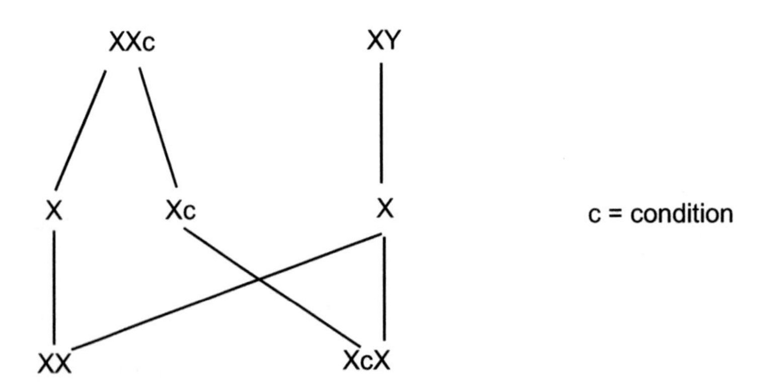

Figure 4.7 *X-linked inheritance 2*

X-linked recessive inheritance

Most changes to the X chromosome in the form of altered or mutated genes are commonly recessive. Men are always affected by a recessive X-linked gene whereas women need both alleles to be in a recessive form before that trait can be expressed. If a woman has one recessive allele and one dominant allele, she is classified as an X-linked carrier for the recessive allele. Males cannot be X-linked carriers as the trait is always expressed.

Characteristics of X-linked recessive inheritance

1. Commonly affects more males than females.
2. Affected males never pass on the affected trait to their sons (no male-to-male transmission of the X chromosome) (see Figure 4.8).

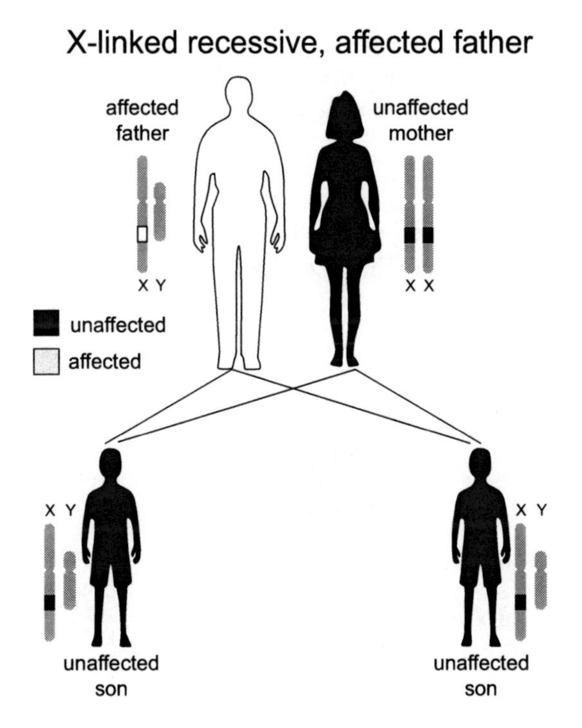

Figure 4.8 *Affected male to sons*

3. Affected males pass on the affected gene to all daughters (see Figure 4.9).

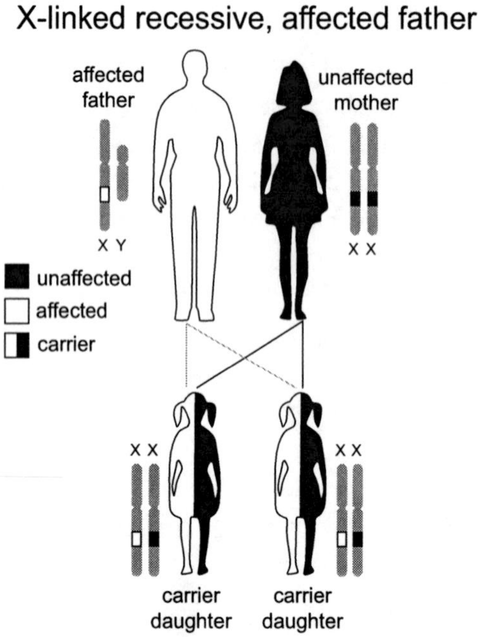

X-linked recessive, affected father

affected father
unaffected mother
X Y
X X
unaffected
affected
carrier
X X
X X
carrier daughter
carrier daughter

Figure 4.9 *Affected male to daughters*

4. Female carriers can pass the defective X chromosome to half of their sons (affected) and half of their daughters (carriers). The other children inherit the unaffected X chromosome (see Figure 4.10).

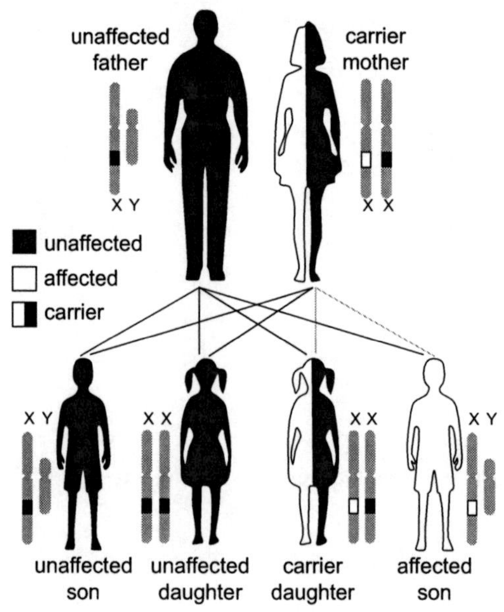

Figure 4.10 *Female carriers to offspring*

5. The overall pattern of inheritance is by transmission of the defective gene from affected males to male grandchildren via a carrier daughter (see Figure 4.11).

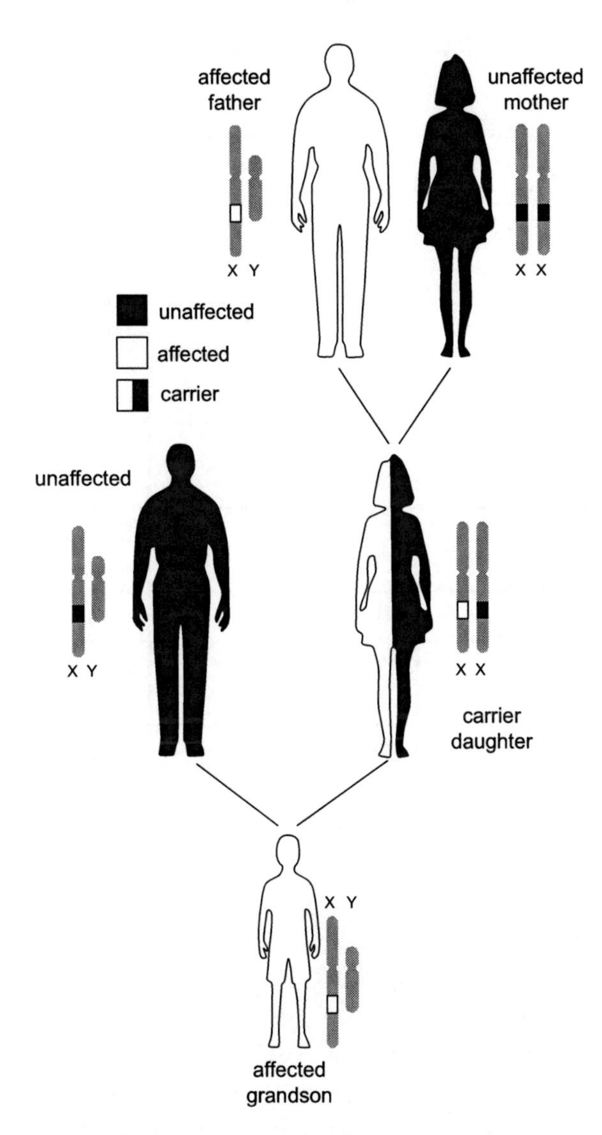

Figure 4.11 *Affected males to grandsons via carrier daughter*

6. Affected females with two deficient alleles are from an affected father and a carrier mother.

Some examples of X-linked recessive disorders are shown in Table 4.1.

Table 4.1 *Examples of X-linked recessive disorders*

Name	Gene	Effects
Alport Syndrome	COL4A5	Hearing loss and kidney failure
Anhidrotic Ectodermal Dysplasia	EDA	Missing sweat glands, hair and teeth
Colour Blindness (red-green)	OPN1LW OPN1MW	Lack of red/green colour perception
Duchenne & Becker Muscular Dystrophy	DMD	Degenerative muscle weakness
Fabry Disease	GLA	Increased lipid storage in endothelial cells leading to heart attacks and kidney failure
Haemophilia	F8 F9	Delayed blood clotting
Menkes Disease	ATP7A	Brain degeneration, abnormal copper transport
Norrie Disease	NDP	Eye degeneration
Severe Combined Immune Deficiency	ADA	Lack of B and T lymphocytes
Wiskott–Aldrich Syndrome	WAS	Infections, rashes and bleeding

ACTIVITY 4.4

Fabry disease

Fabry disease is an enzyme disorder that leads to the accumulation of certain types of lipids (sphingolipids) within the endothelial lining of blood vessels, the myocardium and the glomeruli of the kidney nephrons. Symptoms that often present in childhood include burning sensations in the hands and feet and angiokeratomas (small purple blemishes on the skin). An inability to sweat is also often present, as is corneal cloudiness due to fat deposits within the corneas. Later features include strokes and heart attacks as well as kidney failure.

The disorder often presents with different severities in different families due to over 300 mutations found in the GLA gene on the X chromosome. This gene normally encodes for the enzyme ceramide trihexosidase, which normally metabolises sphingolipids. The altered gene acts in a recessive fashion.

John has Fabry disease but Susan, his wife, does not. There is no history of Fabry disease in Susan's family. What is the chance that any of their children would have Fabry disease?

ACTIVITY 4.5

Red/green colour blindness

Individuals with red/green colour blindness are unable to distinguish between shades of red and green. This is the most common form of colour blindness in Europe, occurring in 8 per cent of males and 0.5 per cent of females. In females who are carriers for this altered gene, one normally functioning gene on the other X chromosome is enough to give full colour vision. The altered gene acts in a recessive fashion.

a. By using **X** and **Y** to indicate the sex chromosomes and the letter **c** for the colour blindness recessive allele, what are the possible genotypes and their corresponding phenotypes in both males and females?

b. Can two colour-blind parents have a son with full colour vision?

c. Can two colour-blind parents have a daughter with full colour vision?

d. Can a daughter with full colour vision have a colour-blind father?

e. Can a son who has full colour vision have a colour-blind mother?

f. A woman (who has full colour vision) has a colour-blind father. This woman marries a colour-blind man and has two children (a son and a daughter) from that marriage. What are the chances that:
  i) the son is colour blind?
  ii) the daughter is colour blind?

X-linked dominant inheritance

Unlike X-linked recessive inheritance, there is no difference in the inheritance risk of X-linked dominant traits between sons and daughters. When a mother carries a dominant gene on the X chromosome then birth sons and daughters have a 50 per cent chance of inheriting this dominant gene. This inheritance from the maternal line is similar to that of autosomal dominant inheritance (see Figure 4.12).

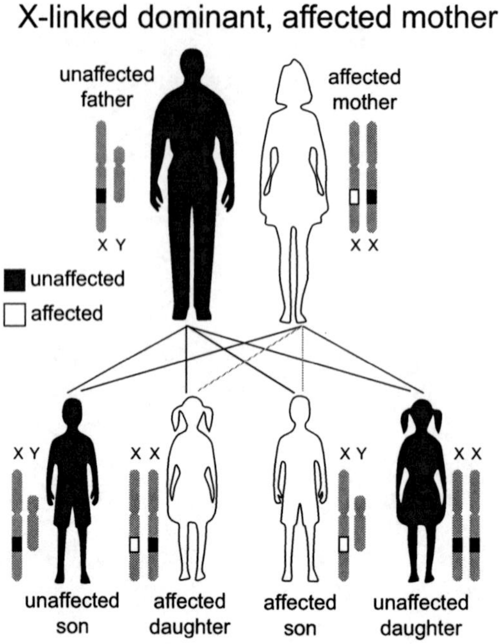

Figure 4.12 *Inheritance from the maternal line*

When a father is affected, the affected gene is passed to all daughters, but not to his sons (see Figure 4.13).

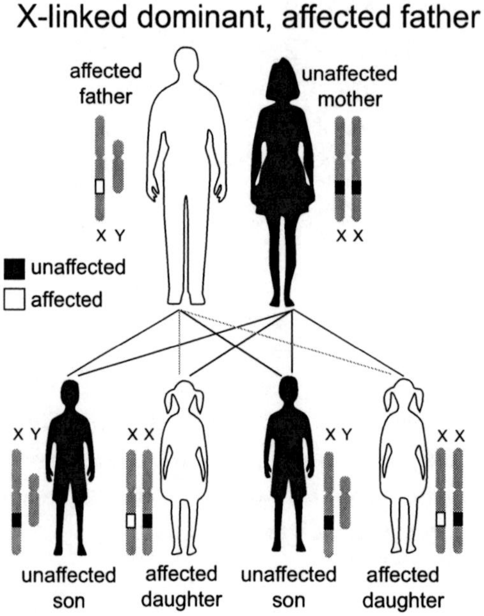

Figure 4.13 *Inheritance from the paternal line*

Characteristics of X-linked dominant inheritance

1. Never passed from father to son.
2. All daughters of an affected father and a non-affected mother are affected.
3. All sons of an affected father and a non-affected mother are not affected.
4. Affected females and non-affected males result in 50 per cent of sons being affected and 50 per cent of daughters being affected.
5. Males are usually more severely affected than females. The trait may even be lethal in males but not in females (due to the effect of not having another X chromosome).
6. More females are likely to be affected (thought to be due to the lack of another X chromosome in males making the affected gene lethal in males).

There are many fewer X-linked dominant conditions in humans in comparison to X-linked recessive conditions. Examples of X-linked dominant disorders are given in Table 4.2.

Table 4.2 *Examples of X-linked dominant disorders*

Name	Gene	Effects
Amelogenesis Imperfecta	AMELX	Abnormal tooth enamel
Congenital Generalised Hypertrichosis	HTC2	Denser and more abundant hair on the upper body
Hypophosphataemia	PHEX	Vitamin D-resistant rickets
Incontinentia Pigmenti	NEMO	Brown pigmented skin on limbs in females, lethal in males
Rett Syndrome	MECP2	Neurodegeneration and reduced cognitive functioning in females, lethal in males

ACTIVITY 4.6

Rett syndrome

Rett syndrome is caused by alterations in the MECP2 gene, which normally encodes for a protein needed for synapse structure as well as the production of other proteins in nerve cells. The affected gene acts in a dominant fashion, but one normal gene is necessary for survival. Males who inherit this altered gene do not survive due to the lack of another X chromosome.

A study by Moog *et al.* (2003) found that, in some rare cases, boys have developed Rett syndrome. Given that an unaffected X chromosome has to be present for survival, what would be the possible sex chromosome genotype for these boys?

The key to identifying if a dominant trait is linked to the X chromosome or an autosomal trait is to examine the family history for any father-to-son transmissions. Any father-to-son transmission would indicate an autosomal involvement. If all daughters are affected but no sons are, then this strongly indicates that it is an X-linked trait.

Y-linked inheritance

Most of the genes that are on the Y chromosome are concerned with sexual function in males. Ninety-five per cent of the Y chromosome does not have corresponding alleles on the X chromosome and is termed the male-specific region. All the genes are expressed as there is only one copy of the Y chromosome.

Characteristics of Y-linked inheritance

1. No females are affected.
2. No 'skipping' of generations.
3. All male offspring are affected.

Y-linked conditions, arising from altered genes on the Y chromosome, are rare as there are so few genes on the Y chromosome. No Y-linked conditions other than infertility have yet been clearly identified.

Pseudoautosomal inheritance

Five per cent of the Y chromosome contains pseudoautosomal genes that pair up with corresponding regions within the X chromosome. Males have two copies of these alleles, one on the X chromosome and one on the Y chromosome. Females can inherit an allele originally present on the Y chromosome of their father, due to the crossing-over of genetic material between the pseudoautosomal regions of the sex chromosomes during meiosis.

There are only a very small amount of genes present on the pseudoautosomal region of the Y chromosome, but these genes are inherited in the same way as any other autosomal genes. Inheritance patterns therefore depend on whether the gene acts in a dominant or recessive fashion.

ACTIVITY 4.7

One of the genes situated on the pseudoautosomal region of the sex chromosomes is the SHOX gene (Short Stature Homeobox). This gene normally encodes for a protein that regulates the activity of other genes. A non-functioning SHOX gene leads to skeletal abnormalities, including shortening of the limbs and unusual joint rotations in the wrists and elbows. The presence of just one affected gene will give rise to this trait.

Would a son of an affected father and a non-affected mother have this condition?

Explain your answer.

X INACTIVATION

The X-linked genes are all paired in females but present as single copies in males. It would be logical to think that the proteins regulated for by the X-linked genes would be doubled in females. This is not so. The amount of protein produced is roughly the same in both males and females. This phenomenon is termed 'dosage compensation' in that a female with two X chromosomes does not 'over-dose' with the X-linked gene products and the male does not 'under-dose'.

The process of dosage compensation is achieved by allowing only one X chromosome to be active in every cell. In females, one of the X chromosomes becomes inactive. X inactivation results in one of the X chromosomes becoming highly condensed and inactive. This condensed X chromosome is referred to as a Barr body. Normal males never express these Barr bodies. The number of Barr bodies in a human cell is one less than the number of X chromosomes within that cell. A normal XX female will have one active X chromosome and one Barr body in every cell.

The Lyon hypothesis

Mary Lyon (1961) proposed the Lyon hypothesis, which states that only one X chromosome is active in any cell. This process of inactivation occurs during early embryonic development. The second X chromosome in females becomes inactivated at the 100 cell stage of development. Which X chromosome is effectively 'turned off' occurs at random, unless one of the chromosomes is damaged (the damaged X chromosome is always 'turned off'). The embryonic tissue in normal development will then consist of some cells with one maternally inherited X chromosome 'turned off' and others with the paternally inherited X chromosome 'turned off'. These inactivated X chromosomes form Barr bodies, and this process is called Lyonisation.

As some female cells will have a maternally active X chromosome and other cells will have a paternally active X chromosome, the female can be described as mosaic. If both the X chromosomes carry homozygous genes, then the X inactivation will have no effect. However, X inactivation of heterozygous genes leads to the expression of the allele on the activated X chromosome, whether the allele is recessive or dominant. A heterozygous female who expresses the recessive gene is termed a 'manifesting heterozygote'.

In theory, X inactivation equals the expression of X-linked genes in the sexes. However, females are mosaic for the X chromosome in that they possess active paternal X chromosomes and active maternal X chromosomes in different cells. Females who are heterozygous for a particular trait will display both the recessive and dominant traits to some extent. For example, with Haemophilia A, a blood clotting disorder, males who carry this gene for altered blood clotting will have delayed blood clotting. Heterozygous females will have some cells with the affected gene that also causes this condition, but will have other cells that are able to encode for the required clotting factor. These females will be mild haemophiliacs, in that they do have delayed clotting times but, due to the presence of the clotting factor gene in some cells, will produce enough clotting factors to make this a mild condition.

CASE STUDY 4.3

Anhidrotic Ectodermal Dysplasia

Clinical features

This condition results in the abnormal development of hair, sweat glands and teeth. Individuals with this condition often experience difficulties in body temperature regulation due to the inability to sweat. Sparse, thin, slow-growing hair is also associated with this condition, as are sparse or missing teeth.

Genetics

Most cases of this condition arise from an alteration to the EDA gene on the X chromosome. The EDA gene normally encodes for a protein that has a vital role in embryonic development. A normal functioning copy of the gene is needed for essential interaction between the ectoderm and mesoderm and the formation of skin, hair, nails and teeth.

The altered gene is inherited in an X-linked recessive pattern. However, a female carrier of this gene will often display some features, but the effects will be patchy, in that some areas of the skin will be normal and other areas will be devoid of sweat glands and hair. This genetic condition is an example of X inactivation.

Exceptions to the Lyon hypothesis

It is now known that not all regions of the X chromosome are inactivated. Research by Carrel and Willard (2005) has indicated that 15 per cent of the genes on the inactive X chromosome are active in all women, with a further 10 per cent being active in some women.

SEX-RELATED GENETIC EFFECTS

Some sex-influenced traits can be inherited on autosomes as well as on the sex chromosomes. A trait that is absolutely limited to one sex is a sex-limited trait, while a trait that is sex-influenced is only expressed as dominant in one sex but as recessive in another.

Sex-limited traits

Sex-limited genes can be X-linked or autosomal. The traits displayed by sex-limited genes are only shown in one sex. Both men and women may inherit these genes but they are only expressed in one of them. For example, coarse facial hair is a sex-limited trait in humans. The gene responsible, although present in both males and females, is only phenotypically expressed in males. Females do not express this gene as they do not produce the hormone necessary for beard growth. However, females can pass this gene down to their sons who can grow beards after puberty due to the presence of the appropriate hormone.

Sex-limited traits are said to have 100 per cent penetrance in one sex and 0 per cent penetrance in the opposite sex.

Sex-influenced traits

The alleles of sex-influenced traits behave differently depending on the individual's sex. These alleles behave as dominant alleles in one sex and as recessive alleles in the opposite sex as, for example, with male patterned baldness. The AR gene for male patterned baldness is situated on chromosome 15. There are two types of alleles for this gene, one that produces hair on the head and one that causes baldness. The allele that causes baldness acts as an autosomal dominant in men but recessive in women. This gene has also been implicated in polycystic ovary disease in women. Females who have this condition experience a decrease in fertility, irregular menstruation, excess body hair and weight gain. This gene appears to act as an autosomal dominant for ovarian disease, but as an autosomal recessive for hair loss in women.

SUMMARY

- There are two sex chromosomes in the human genome, XX in females, XY in males.
- The Y chromosome is responsible for male differentiation. The absence of a Y chromosome leads to female differentiation.
- Females can only pass an X chromosome to their offspring; males can pass either an X or a Y chromosome. Males determine the sex of their offspring.
- Sex ratio is the number of males in a population divided by the number of females, multiplied by 1,000. Sex ratios are affected by biological, environmental and cultural influences.
- X-linked traits may be dominant or recessive, but are always expressed in males (due to the absence of another X chromosome). X-linked traits are never passed from father to son. Y-linked traits are rare and are only passed from father to son.
- Dosage compensation mechanisms tend to limit the expression of X-linked genes in females. X inactivation results in the formation of Barr bodies by a process called Lyonisation.
- X inactivation occurs at random, resulting in females developing as genetic mosaics for X-linked heterozygous alleles.

- Sex-limited and sex-influenced traits can be sex linked or autosomal. Sex-limited traits only affect one sex and not the other. Sex-influenced genes act as dominant genes in one sex and as recessive genes in the opposite sex.

FURTHER READING

Suggested reading on Asia's missing women is the following articles:

Gupta, M.D. (2005) 'Explaining Asia's "missing women": A new look at the data'. *Population and Development Review* 31(3): 529–35

Jha, P., Kumar, P., Vasa, N., Dhingra, D., Thiruchelvam, R. and Moineddin, R. (2006) 'Low male-to-female sex ratio of children born in India: national survey of 1.1 million households'. *The Lancet* 367(9506): 211–18

Sheth, S.S. (2006) 'Missing female births in India'. *The Lancet* 367 (9506): 185–6

A good source that gives up-to-date statistics on sex ratio within different countries can be found on the Central Intelligence Agency website. This includes *The World Factbook* that gives the latest statistics on sex ratio: **https://www.cia.gov/library/publications/the-world-factbook/index.html**

Suggested reading on X chromosome inactivation includes the following two key papers:

Carrel, L. and Willard, H.F. (2005) 'X-inactivation profile reveals extensive variability in X-linked gene expression in females'. *Nature* 434: 400–4

Lyon, M.F. (1961) 'Sex chromatin and gene action in mammalian X-chromosome'. *American Journal of Human Genetics* 14(2): 135–48

For further reading on the conditions mentioned in this chapter see:

Moog, U., Smeets, E.E.J., van Roozendal, K.E.P., Schoenmakers, S., Herbergs, J., Schoonbrood-Lenssen, A.M.J. and Schrander-Stumpel, C.T.R.M. (2003) 'Neurodevelopmental disorders in males related to the gene causing Rett syndrome in females (MECP2)'. *European Journal of Paediatric Neurology* 7: 5–12

The Online Mendelian Inheritance in Man website is hosted by the Johns Hopkins University in America. It contains detailed information on genetic conditions with good access to relevant research papers:
www.ncbi.nlm.nih.gov/omim

05

TWO OR MORE GENES

LEARNING OUTCOMES

The following topics are covered in this chapter:

- monogenic inheritance;
- probabilities;
- polygenic inheritance;
- multifactorial inheritance;
- common multifactorial conditions.

INTRODUCTION

Nearly all single-gene traits are inherited in a Mendelian fashion. When one gene encodes for one trait it is referred to as **monogenic**. When a human trait is encoded for by more than one gene it is referred to as **polygenic** (this is not to be confused with pleiotropy, which is when one gene can give rise to different traits) (see Figure 5.1).

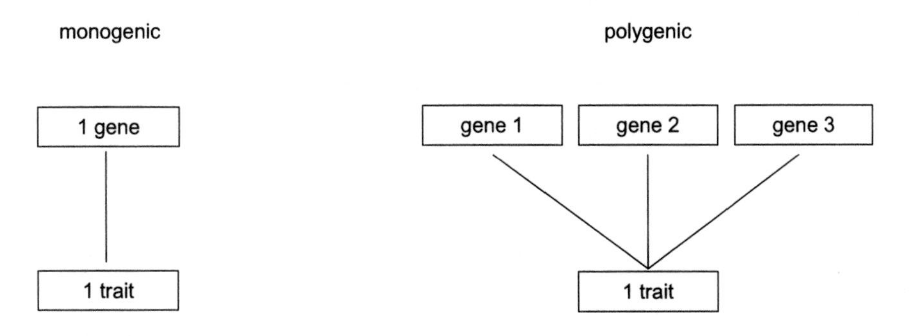

Figure 5.1 *Monogenic and polygenic traits*

Monogenic traits are inherited in a Mendelian inheritance pattern and polygenic traits usually involve complex inheritance patterns.

MONOGENIC INHERITANCE

Mendel's principle of independent assortment states that different genes that control different phenotypic traits reassort independently from each other. This principle only applies to single genes that control a single trait. For example, the gene that encodes for tongue rolling in humans and the gene that encodes for attached ear lobes assort independently from each other. Not all individuals who are able to form a U shape with their tongues have attached ear lobes, because they are different traits encoded for by unrelated genes.

Working out the inheritance pattern of more than one trait can also be done through the construction of a Punnet square. Consider the above example of tongue rolling (a dominant trait) and attached earlobes (a recessive trait).

Trait 1		Trait 2	
Tongue rolling	T	Free ear lobe	E
Non-tongue rolling	t	Attached ear lobe	e

The genotype of a heterozygous individual for tongue rolling and ear lobe form would be TtEe and they would have the ability to tongue roll and have a free ear lobe phenotype. A mating between two heterozygous individuals could give rise to four different gametes from each parent: TE, Te, tE, te.

The Punnet square needs to be large enough to accommodate all the possible gamete formations (Figure 5.2).

gametes

gametes	TE	Te	tE	te
TE	TTEE	TTEe	TtEE	TtEe
Te	TTEe	TTee	TtEe	Ttee
tE	TtEE	TtEe	ttEE	ttEe
te	TtEe	Ttee	ttEe	ttee

Figure 5.2 *A Punnet square of gamete formations*

When two traits are considered, four different gametes are produced in equal frequencies in both the male and the female. A Punnet square shows the result of all 16 possible combinations of those gametes. Due to dominance there are four possible phenotypes in the ratio of 9:3:3:1 (a total of 16). Therefore, the chances of producing:

- a tongue-rolling child with free earlobes (T-E-) is $^{9}/_{16}$
- a tongue-rolling child with attached earlobes (T-ee) is $^{3}/_{16}$

- a non-tongue-rolling child with free earlobes (ttE-) is $^3/_{16}$
- a non-tongue-rolling child with attached earlobes (ttee) is $^1/_{16}$

Punnet squares can be used for more than two monogenic traits, but this method becomes laborious and time consuming, with opportunities for errors occurring. A simpler method to work out the chances of the combined traits in an offspring is by using basic probability mathematics.

Working out the probabilities

Probability is a theoretical concept based on the likely outcome of an event; it does not rely on the event having already taken place. There are two principles used when calculating genetic risks: the law of addition and the law of multiplication.

The law of addition

The probability of one event or another occurring is calculated by **adding** the individual probabilities. For example, the probability for a pregnant woman having a boy or a girl is:

Boy		Girl		
½	+	½	=	1

Addition is only used with events that are related to each other in an 'either/or' format.

The law of multiplication

With two independent events the individual probabilities are multiplied. For example, if a couple are planning on having two children, the probability of both children being girls is:

Girl		Girl		
½	×	½	=	¼

Probabilities are added together when they relate to mutually exclusive alternative outcomes. These are probabilities that are joined together by the word 'or'. Probabilities are multiplied if they relate to independent events. These are usually joined by the word 'and'.

Due to the independent assortment of most genes, the probability of inheriting one trait does not affect the probability of inheriting another different trait. Consider the following three recessive traits: widow's peak hairline, cleft chin and freckles. If both parents were heterozygous for all three traits (carriers but not exhibiting all the traits), the probability of a child being homozygous recessive for all three traits would be as shown in Figure 5.3.

As these traits are independent of each other, to find out the probability of having a child with all three traits the sum would be:

$$\frac{1}{4} \times \frac{1}{4} \times \frac{1}{4} = \frac{1}{64}$$

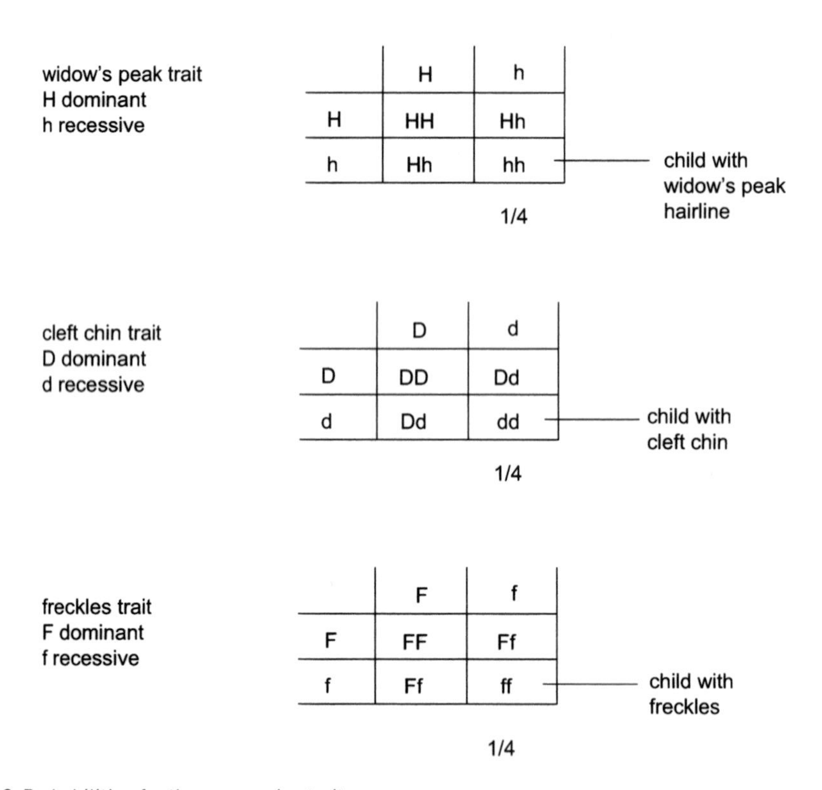

Figure 5.3 *Probabilities for three recessive traits*

Therefore, as all the genes reassort independently during meiosis, the chance of having a child with a widow's peak and a cleft chin and freckles would be 1 in 64. The individual risk measurements are multiplied to give the overall risk of these traits appearing together in one offspring.

The major drawback with calculating the probabilities of inheritance in humans is that it is mainly theoretical. Predictions of outcomes work well with large sample sizes, but offspring numbers in families are not large. This means that the risk of 1 in 4 for having a child with just one of these traits does not mean that the parents have to have four children before one of these traits appears in their offspring.

ACTIVITY 5.1

a. What is the probability that, if a couple had four children, all of the children would be girls?

b. A couple have four boys and desperately want a daughter. What is the probability that the next child will be a daughter?

c. Cystic fibrosis is a monogenic recessive disorder. A couple who are both carriers of the cystic fibrosis gene have two children. What is the chance that both children will be affected by cystic fibrosis?

d. A mother has the genotype AaBbCc and the father is AaBbcc. What is the probability that their offspring will be AaBBCc? (Note: work out each trait separately first then use the rule of multiplication.)

e. Phenylketonuria is a recessive metabolic condition. If both parents are heterozygous carriers of this condition, what is the chance that their five children will be:

i) all normal;
ii) 4 normal/1 affected;
iii) 3 normal/2 affected;
iv) 2 normal/3 affected;
v) 1 normal/4 affected;
vi) all affected.

POLYGENIC INHERITANCE

The one gene–one trait effects described so far follow the Mendelian principles of inheritance. Single gene disorders in humans tend to give rise to a dichotomous (either/or) disorder, in that if the specific gene is present then the disorder is also present. These monogenic disorders are the exceptions rather than the rule. Rarely do single genes control one trait. Most monogenic disorders in humans are quite rare, occurring once in 5,000 to 100,000 individuals.

Most traits are polygenic, in that they are encoded for by a number of different genes that work together. The genetic effect of the different genes encoding for one trait tends to be additive.The additive effect leads to a continuous variation rather than the either/or scenario of the single gene traits. For example, height is a trait that is encoded for by a number of different genes. If height was controlled by a single gene it would result in discontinuous variation, i.e. either short stature or a tall stature, but no intermediate height. Height displays continuous variation, in that people can be any height ranging from short to tall. Many human traits display continuous variation, including eye colour, skin colour, weight, blood pressure, metabolic rate and even behaviour.

The words 'continuous' and 'discontinuous' are used to describe phenotypic and not genotypic variation. Continuous characters are polygenic. The number of genes that control a trait does not have to be very large before the phenotypic variation can be described as continuous.

Eye colour in humans is known to be a polygenic trait as there is continuous variation of eye colour within populations. Imagine if eye colour was controlled by two genes, this would give rise to five different phenotypes (Figure 5.4):

	gametes			
gametes	AB	Ab	aB	ab
AB	AABB	AABb	AaBB	AaBb
Ab	AABb	AAbb	AaBb	Aabb
aB	AaBB	AaBb	aaBB	aaBb
ab	AaBb	Aabb	aaBb	aabb

Figure 5.4 *Eye colour genes: Aa Bb*

Phenotypes

4 dominant genes	dark brown eyes
3 dominant genes	hazel eyes
2 dominant genes	green eyes
1 dominant gene	dark blue eyes
0 dominant genes	light blue eyes

The additive effect of these genes working together can be seen by the fact that the presence of dominant alleles (whether from gene A or B) results in darker colouration, so an individual who has the genotype AABB has very dark eyes and the individual who possesses only recessive genes (aabb) has the lightest eye colouring of all.

If three genes have an additive effect towards one trait then the possible phenotypic variation is enlarged. Three genes can result in seven different phenotypes. Human skin colour is a polygenic trait, but the number of genes responsible for skin pigmentation remains unclear. Imagine: if this trait was encoded for by three genes, Aa, Bb and Cc, this would result in seven different shades of pigmentation (Figure 5.5).

The variations in skin colour in humans cannot be categorised into seven different shades, which suggests that the number of genes responsible for skin colour is larger than three.

Most polygenic traits that show continuous variation are measurable like, for example, height, skin colour and weight. These measurable traits are also known as quantitative traits. Human polygenic traits:

- are due to the action of two or more genes;
- have a measurable phenotype;
- have a wide range of variable phenotypic expression.

gametes	abc	Abc	aBc	abC	ABc	AbC	aBC	ABC
abc	aabbcc	Aabbcc	aaBbcc	aabbCc	AaBbcc	AabbCc	aaBbCc	AaBbCc
Abc	Aabbcc	AAbbcc	AaBbcc	AabbCc	AABbcc	AAbbCc	AaBbCc	AABbCc
aBc	aaBbcc	AaBbcc	aaBBcc	aaBbCc	AaBBcc	AaBbCc	aaBBCc	AaBBCc
abC	aabbCc	AabbCc	aaBbCc	aabbCC	AaBbCc	AabbCC	aaBbCC	AaBbCC
ABc	AaBbcc	AABbcc	AaBBcc	AaBbCc	AABBcc	AABbCc	AaBBCc	AABBCc
AbC	AabbCc	AAbbCc	AaBbCc	AabbCC	AABbCc	AabbCC	AaBbCC	AABbCC
aBC	aaBbCc	AaBbCc	aaBBCc	aaBbCC	AaBBCc	AaBbCC	aaBBCC	AaBBCC
ABC	AaBbCc	AABbCc	AaBBCc	AaBbCC	AABBCc	AABbCC	AaBBCC	AABBCC

phenotypes:							
	1/64	6/64	15/64	20/64	15/64	6/64	1/64
number of dominant alleles:	0	1	2	3	4	5	6

Figure 5.5 *Seven different shades of pigmentation*

ACTIVITY 5.2

Why are some polygenic traits considered quantitative?

Polygenic disorders

In contrast to single-gene disorders, polygenic disorders result from less severe alterations in more than one gene (Table 5.1). Any one of these alterations on their own might not affect the phenotype but, together, they can lead to a significant alteration in the phenotype. Inheritance of most traits are quite complex because they are influenced by multiple genes, most of which have pleiotropic effects (see Figure 5.6). Pleiotropy reflects the fact that single proteins have multiple roles within the body.

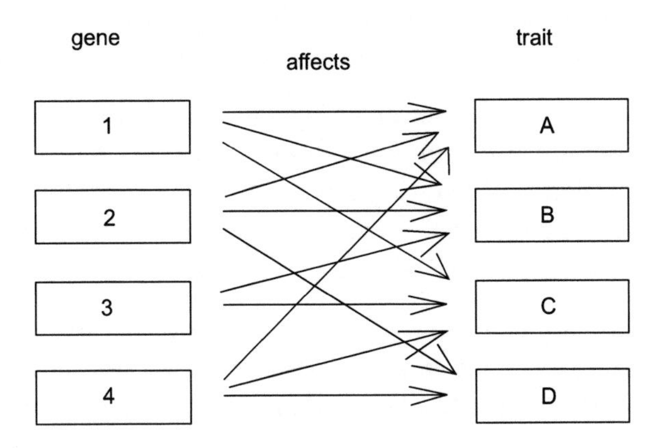

Figure 5.6 *Complex traits influenced by multiple genes*

Table 5.1 *Human monogenic and polygenic genetic disorders*

	Monogenic disorders	Polygenic disorders
Examples	Cystic Fibrosis Huntington's Disease PKU Duchenne Muscular Dystrophy Fragile X Syndrome Haemophilia Adult Polycystic Disease	Heart Disease Hypertension Alzheimer's Disease Diabetes Cancer Arthritis Eczema Bipolar Disorder Schizophrenia
Frequency	Uncommon to extremely rare	Often common
Inheritance Patterns	Simple and well understood	Complex, inconsistent and poorly understood
Genetic Tests	Not available for all single-gene disorders Valuable for affected families Helpful for decision making	Poor predictive value Limited use

MULTIFACTORIAL INHERITANCE

Genes rarely act alone. Both monogenic and polygenic traits can also be multifactorial. Multifactorial traits arise from both genetic and environmental factors. Most polygenic traits are multifactorial traits. The phenotypes of multifactorial traits are influenced by environmental factors. Polygenic traits that are not multifactorial are very rare.

Height is a multifactorial trait. An individual might have the genetic make-up to become tall but, without an adequate diet during childhood development, the full height potential might never be achieved. Skin colour can be affected by the environmental influence of the sun, and the lack of sufficient levels of oxygen at birth can have a lifelong effect on learning abilities.

Multifactorial traits are also known as complex traits as the inheritance pattern in families can be difficult to identify. Although the individual genes are inherited in a Mendelian fashion, polygenic multifactorial traits might involve a number of different genes, which do not all have an equal input towards the resulting phenotype. Each gene confers a degree of susceptibility that results in an additive effect.

The patterns of inheritance of multifactorial traits are less predictable than those of purely genetic traits. The genes of a multifactorial condition make the individual susceptible to that condition, but other environmental factors need to be present for the condition to manifest.

Most of the common disorders of adults are multifactorial. Coronary heart disease, hypertension, diabetes, Alzheimer's Disease, schizophrenia, bipolar disorder and some cancers are multifactorial conditions affecting adults, while cleft lip and palate and spina bifida are two examples of congenital multifactorial disorders.

The genes in a multifactorial trait tend to have variable penetrance as the condition does not always develop, despite the genes being present. Incomplete penetrance is due to the interaction of the genetic information and the environment. It is the incomplete penetrance that makes the detection and tracing of genetic disorders in families difficult. The risk for an individual in developing such a condition is estimated by:

- examining the family history;
- having one or more affected biological relatives (usually below the age of 50 years);
- how closely related the affected relative is;
- in some conditions, how early the symptoms first appear.

A condition that 'runs' in the family can be as a result of shared environmental factors such as smoking, poor diet and lack of exercise.

ACTIVITY 5.3

What is the difference between a monogenic trait, a polygenic trait and a multifactorial trait?

Working out the inheritance risks

Many polygenic multifactorial conditions have poorly understood inheritance patterns. This makes it difficult to predict the risk of developing a condition that 'runs' in the family. In these circumstances the **empiric risk** is used. This is not a calculation as such but a population statistic, based on the incidence (rate of occurrence) of the condition within a population and the prevalence (number of individuals who have the condition at that time) within the same population. The empiric risk increases with both the severity of the disorder and the closeness of the affected relative.

Table 5.2 *The empiric risk of schizophrenia*
Source: Scourfield and McGuffin (1999)

Relationships	Empiric risk
Identical twin	46%
Sibling	9%
Child	15%
Niece/nephew	5%
First cousin	3%
General population	1%

In order to determine how much influence the genes have in the development of a multifactorial trait a measure of **heritability** is used. Heritability estimates how much the genes contribute towards a particular condition or trait. The environmental influences can also be estimated once the heritability has been established. For a single dominant gene trait that has 100 per cent heritability, such as Huntington's disease, the heritability is measured as 1.0. For the multifactorial trait of height, the heritability is measured as 0.8, which indicates that height is 80 per cent genetic and 20 per cent environmental. The heritability applies to populations and not to individuals. The heritability of some common conditions is shown in Table 5.3.

Table 5.3 *Heritability of some common conditions*

Condition	**Heritability**
Asthma	0.8
Alzheimer's Disease	0.8
Bipolar Disorder	0.7
High Blood Pressure	0.6
Body Mass Index	0.5
Cleft lip/Palate	0.75
Club Foot	0.8
Coronary Heart Disease	0.65
Height	0.8
Myopia	0.9
Neural Tube Defect	0.6
Peptic Ulcer	0.4
Schizophrenia	0.8
Serum Cholesterol	0.6

As heritability measures apply to populations, the heritability measures can vary between different ethnic groups due to the restriction of gene pools, as well as between different cultures.

ACTIVITY 5.4

a. Explain the difference between heritability and empiric risk.

b. Asthma affects up to 1 in 7 children in some societies and 15 million worldwide. It is an inflammatory disease of the small airways of the lungs characterised by intermittent narrowing of the airways causing airflow obstruction. Heritability ranges from 36 per cent to 75 per cent depending on which population is sampled. Why might heritability results vary in different populations?

COMMON MULTIFACTORIAL DISORDERS

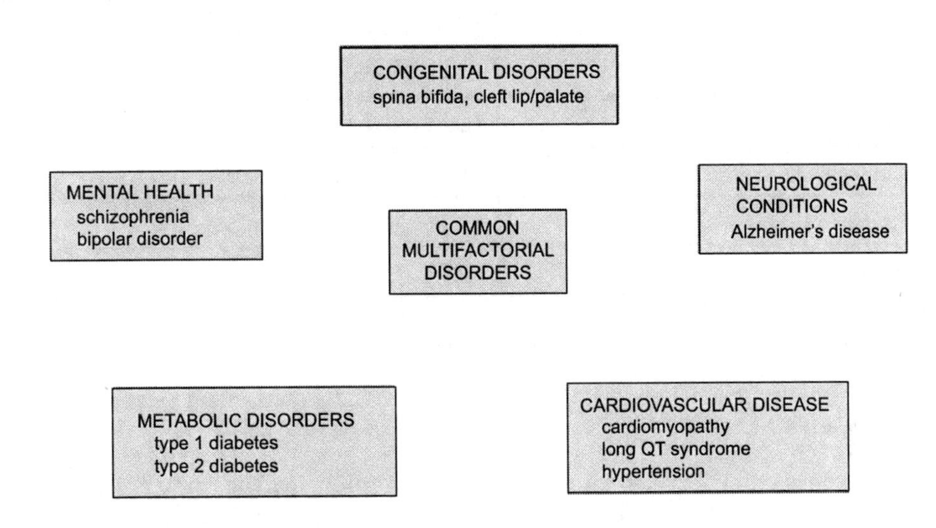

Figure 5.7 *Common multifactorial disorders*

Metabolic disorders

Diabetes Mellitus

Diabetes is a condition that is characterised by high levels of blood glucose (hyperglycaemia). These high levels are a result of the body's failure to produce enough of the insulin hormone or the lack of response by the body's cells to the insulin produced. Insulin is produced by the pancreas and is needed for the uptake of glucose into body cells so that glucose can then be converted into energy. There are two main forms of diabetes mellitus: type 1 and type 2.

Type 1 and type 2 diabetes are different conditions, with different causes and different genetics (Table 5.4). Type 1 is caused by an autoimmune destruction of the insulin-producing beta cells within the pancreas. This results in the affected individual requiring life-long insulin replacement therapy. Type 1 diabetes usually develops during childhood. Type 2 diabetes is usually caused by a combined result of reduced insulin secretion and cellular unresponsiveness for the uptake of glucose. Environmental factors such as age, obesity and lack of exercise are identified risk factors associated with Type 2 development.

Table 5.4 *A comparison of type 1 and type 2 diabetes*

Type 1	**Type 2**
10% of all cases	90% of all cases
Juvenile onset	Maturity onset
0.4% of UK population	2% of UK population
20 million worldwide	220 million worldwide
Requires insulin therapy	Insulin not always required
No link to obesity	Linked to obesity
Empirical risk	**Empirical risk**
Identical twin: 30% to 50%	Identical twin: 50% to 100%
Sibling: 6% to 10%	Sibling: 30%
Genetics	**Genetics**
HLA genes on chromosome 6	No association with the HLA genes

Type 1 diabetes

The HLA genes on chromosome 6p modulate the immune defence system of the body. Over 90 per cent of individuals with type 1 diabetes have altered HLA genes. Other genes have also been studied such as, for example, the IDDM2 gene on chromosome 11p. This gene is closely positioned to the structural gene for insulin and is thought to be involved in some type 1 cases, as is the CTLA4 gene on chromosome 2q, which has been associated with autoimmune disease.

The increasing evidence of type 1 diabetes suggests that environmental factors may be involved. Early exposure to some viruses has been suggested as an environmental trigger but this has yet to be confirmed.

Type 2 diabetes

This is a relatively common condition in affluent countries and it has long been known that it is, in part, inherited. The disease itself may be present in a sub-clinical form long before a diagnosis has been made. The onset may be hastened by stress, drugs or illness. The four major risk factors for developing type 2 diabetes are increasing age, ethnicity, family history and obesity (especially excess weight around the waist area of the body). The prevalence in the UK is around 2 per cent and first degree relatives are three times more likely to develop type 2 diabetes than individuals who have no family history. This confirms a strong genetic influence but, to date, the genes responsible for the majority of cases have yet to be identified. However, research is now being focused on a number of susceptibility genes.

An association has been made between low birth weight and glucose intolerance in later life. This is thought to be due to the poor development of the beta cells within the pancreas, which would predispose the individual to develop type 2 diabetes in later life. There is no evidence of an autoimmune involvement in type 2 diabetes.

Cardiovascular disease

Although there are some monogenic disorders that give rise to cardiovascular disease, most adult onset cardiovascular disease is polygenic and multifactorial. Polygenic disorders such as hypertension and lipid abnormalities, and environmental influences such as smoking, obesity and lack of exercise, all contribute to cardiovascular disease. In most cases where there is a family history of cardiovascular disease, the cause is multifactorial. The genetics govern the individual's susceptibility rather than the cause of the disease itself.

Cardiomyopathies

Cardiomyopathy refers to an abnormality within the muscle of the heart. There are many different types of cardiomyopathies but two types are known to have a genetic basis.

1. Familial Hypertrophic Cardiomyopathy

This is where the wall of the ventricles of the heart (particularly the left) has enlarged. The muscle thickens and alters the heart structure, which can then lead to palpitations, chest pain and breathlessness. This condition has a variable age of onset, ranging from birth up to the tenth decade of life. Hypertrophic cardiomyopathy is caused by defective sarcomeres. The sarcomere is the basic contractile unit of the cardiac muscle. To date, 12 genes have been identified that encode for different components of the sarcomere. More than 900 different mutations or alterations have been identified with these 12 genes. However, alterations in the MYH7 gene on chromosome 14q and the MYBPC3 gene on chromosome 11p account for over 40 per cent of familial hypertrophic cardiomyopathy cases. Mutations in any of the twelve sarcomere encoding genes can be inherited in an autosomal dominant fashion. A child with an affected parent has a 50 per cent risk of inheriting the mutated gene and a 90 per cent risk of developing the condition. The causes of the mutations in the genes that lead to this condition have yet to be identified.

2. Familial Dilated Cardiomyopathy

Dilated cardiomyopathy refers to the enlargement of the heart ventricles and the thinning of the ventricular walls. This leads to poor contraction. Up to 50 per cent of dilated cardiomyopathy cases are familial, in that at least two members of the family have the same condition. The age of onset is highly variable, ranging from birth to old age. It is estimated that 20 to 50 per cent of dilated cardiomyopathy cases have a genetic component, and current research shows that over 20 genes may be involved with the development of the disorder. The LMNA gene on chromosome 1q affects 7 per cent to 8 per cent of cases and the MYH7 gene on chromosome 14q affects 5 per cent to 8 per cent of cases. The risk of an offspring developing the same condition as one of the parents is 50 per cent. The risk of a sibling of an affected brother or sister (with unaffected parents) is low but still higher than the normal population. No environmental factors have been identified as yet with familial dilated cardiomyopathy.

Long QT Syndrome

Long QT Syndrome is a group of conditions that is characterised by prolonged QT intervals on an electrocardiograph. The heart muscle takes longer than usual to recharge between beats and the affected individual has an irregular heart beat. This can lead to recurrent fainting episodes (syncopes), cardiac arrest and sudden death. There are at least 12 different forms of long QT syndromes, the most common being Romano–Ward Syndrome and Jervell–Lange Nielsen Syndrome.

1. Romano–Ward Syndrome

This is the most common form of the disorder, affecting 1 in 7,000 individuals (this may be a conservative estimate as many go undiagnosed). The Romano–Ward Syndrome is inherited in an autosomal dominant fashion and is caused by defective alterations in a number of different genes that encode for cellular membrane channels (see Table 5.5). These channels are needed for the movement of sodium and potassium ions in and out of the cells. Disruption to the ion movement alters the heart beat leading to prolonged QT intervals and arrhythmias. Mutations in any of the genes in Table 5.5 can disrupt ion flow.

Table 5.5 *Genes identified in Romano–Ward Syndrome*

Gene	Chromosome
KCNE1	21q
KCNE2	21q
KCNH2	7q
KCNQ1	11p
SCN5A	3p

Some medications may trigger the condition, as will strenuous exercise in some individuals. Low potassium levels have also been found to be a trigger for this condition.

2. Jervell–Lange Nielsen Syndrome

This is a rarer form of Long QT Syndrome in comparison with Romano–Ward Syndrome. About 3 people in 1 million are affected with this disorder worldwide, and it is often accompanied by congenital deafness. Alterations in both KCNQ1 gene on chromosome 11p and the KCNE1 gene on chromosome 21q have been linked with this syndrome. Having just one faulty copy of either of the identified genes in any Long QT Syndrome is not enough for the individual to develop the disorder, but it does make the individual susceptible.

Hypertension

Elevated arterial blood pressure is a major cause of vascular disease, which can lead to cerebral haemorrhage, coronary artery disease and renal failure. Blood pressure is a characteristic of each individual with marked variation. It is expressed in two measurements: the systolic measurement represents the force exerted by the blood against the arterial walls when the heart contracts; the diastolic measurement represents the blood force against the arterial walls between heart contractions.

Hypertension is categorised as an elevated systolic pressure of over 140 mmHg and an elevated diastolic pressure of over 90 mmHg (normal is approximately 120 mmHg systolic and 80 mmHg diastolic) (see Table 5.6).

Table 5.6 *Classification of blood pressure levels (British Hypertension Society, 2003)*

Category	**Systolic**	**Diastolic**
Blood Pressure		
Optimal	< 120 mmHg	< 80 mmHg
Normal	< 130 mmHg	< 85 mmHg
High Normal	130–139 mmHg	85–89 mmHg
Hypertension		
Mild	140–159 mmHg	90–99 mmHg
Moderate	160–179 mmHg	100–109 mmHg
Severe	≥ 180 mmHg	≥ 110 mmHg

Most cases of hypertension (80–90 per cent) have no obvious cause and are termed 'essential hypertension'. Essential hypertension frequently occurs in family clusters, and individuals who have two or more affected relatives have a three times increased risk of developing hypertension. The cause is not likely to be a change in a single gene but probably involves the interaction of numerous altered genes. Biochemical studies have implicated both the angiotensin I converting enzyme (ACE) and the angiotensin II type 1 receptor, which are both part of the renin–angiotensin homeostatic control mechanism for blood pressure, as probable genetic causes of hypertension (see Figure 5.8).

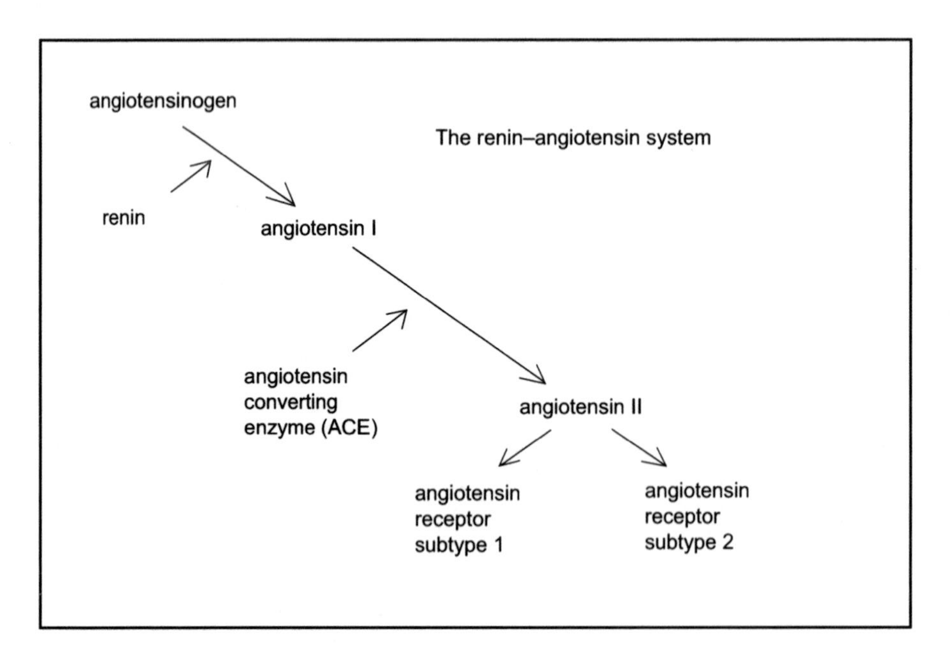

Figure 5.8 *The Renin–Angiotensin System*

Variations in the gene that encodes for angiotensinogen have been linked to hypertension in both European and Japanese studies but it has not been demonstrated in African-Americans. The angiotensin-converting enzyme (ACE) converts angiotensin I to angiotensin II, which acts as a powerful vasoconstrictor. The ACE gene on chromosome 17q has been discovered in an altered form in various different studies, with some altered forms associated with increased serum ACE activity.

Angiotensin receptor subtype 1 mediates the vasoconstrictive effects of angiotensin II and the release of aldosterone (which, in turn, increases the sodium retention in blood plasma). Changes to the gene that encodes for the angiotensin receptor subtype 1 have been found in a Finnish study but again not in African-American studies. Inconsistent findings are probably as a result of small-scale studies, variable susceptibilities in different populations and the combined effect of numerous genes.

Environmental factors such as a high sodium diet, smoking and lack of exercise all contribute towards the development of hypertension.

Mental health disorders

Mental health is an umbrella term that encompasses disorders of the mind. Mental health conditions include depression, anxiety, schizophrenia, bipolar disorder and personality disorders. Most of the genetic research in mental health has been on the conditions of schizophrenia and bipolar disorder. It is these two conditions that will be outlined here. Much of the evidence for a genetic basis in mental health has arisen from studies of twins. Identical twins who have been raised apart, as well as adopted children of affected parents, have given researchers the opportunities to assess the genetic and environmental influences on mental health.

1. Schizophrenia

Schizophrenia is an umbrella term for a number of different conditions that display a wide variety of symptoms. Not all individuals diagnosed with the condition will exhibit all of the symptoms. The disorder is a psychosis, a disorder of thought, losing touch with reality, hallucinations, delusions and paranoia. Although mood is often affected, it is distinguished from mood disorders in which affected mood is a primary symptom. There may also be impairment of cognitive function but it is distinguished from dementia where cognitive decline is a primary symptom.

The symptoms of schizophrenia usually occur in late adolescence or early adulthood and the risk of developing schizophrenia for the general population is 1 per cent. This risk increases if there is an affected family member. This strongly suggests a genetic component to the disorder (see Table 5.7).

Table 5.7 *The empiric risk of schizophrenia*

Empiric risk of schizophrenia	%
Identical twin	46
Both parents	45
One sibling and one parent	15
One parent	11
One sibling	9
Aunt/uncle/grandparent	4
First cousin	3
General population	1

The genetic component of schizophrenia encompasses many different genes that increase the individual's susceptibility to the disorder. It is likely that these different genes may be involved in increasing the susceptibility of developing different forms of schizophrenia. Numerous research studies have linked schizophrenia with gene changes on chromosomes 1q, 1p, 2q, 5p, 5q, 6q, 8p, 10q, 11q, 13q, 15q and 22q. Changes in the dysbindin gene on chromosome 6p, the neuroregulin gene on 8p, the G72 gene on chromosome 13q and five different genes on chromosome 22q have all been identified and associated with the development of schizophrenia.

There appears to be an increased risk of schizophrenia in an offspring if the father was over the age of 50 years at the time of conception. However, the environmental factors for schizophrenia remain largely unidentified. Some research suggests that complications during pregnancy and delivery and the use of some recreational drugs such as cannabis and methamphetamine may act as environmental risk factors.

2. Bipolar disorder

Bipolar disorder is characterised by extreme mood swings ranging from mania to depression. The mood swings usually appear to occur spontaneously, without any apparent external cause.These extreme mood swings can be interspersed with periods of 'normal' mood but, in some individuals, the depression and mania alternate rapidly. The disorder has been subdivided into two main categories: bipolar 1 and bipolar 2.

Bipolar 1

This is the classic form of the disorder. Individuals with type 1 bipolar disorder will develop extreme forms of mania in which they can become psychotic, develop delusions and experience hallucinations.

Bipolar 2

This condition is characterised by milder forms of mania and, although judgement can be affected, the individual does not become psychotic and they will not experience delusions or hallucinations.

The main difference between type 1 and type 2 relates to the manic state of the disorder, whereas there is no difference between the two types in the depressive phase. Onset is highly variable for both types, ranging from pre-school to old age.

Twin, family and adoption studies strongly point to a genetic basis for bipolar disorder. Genetic studies have suggested many different probable genes, although many of these studies are inconclusive, inconsistent and often not replicated. Some of the genes thought to be linked to bipolar disorder include the FAT gene on chromosome 4q, the CUX2 gene on chromosome 12q and changes to a region within chromosome 18q have also been associated with bipolar disorder type 2 (see Table 5.8).

Environmental factors play a significant role in the development of bipolar disorder. Studies have shown that between 30 and 50 per cent of adults with the condition reported traumatic and abusive events in their childhood. However, other environmental factors affecting the development of bipolar disorder remain largely unidentified.

Table 5.8 *The empiric risk for bipolar disorder*

Empiric risk for bipolar disorder	%
Identical twin	70
Both parents	50
One sibling and one parent	20
One parent	15
One sibling	13
Aunt/uncle/grandparent	5
Cousin	3
General population	3

Neurological disorders

Alzheimer's disease

Alzheimer's disease is a form of dementia that causes gradual loss of memory and judgement, which leads to functional disability. This disorder is a degenerative disease of the brain that results in cerebral atrophy, interneuronal neurofibrillary tangles and beta amyloid plaque formation. Affected individuals might also suffer from language disturbances, agitation, hallucinations, withdrawal, decreased muscle tone, incontinence and even seizures. Total care is required in the advanced stages of the disease. Death usually occurs as a result of pneumonia, malnutrition or general body wasting.

The prevalence of Alzheimer's disease in older populations is quite high, affecting 5 per cent of over 70 year olds, increasing to 25 per cent to 45 per cent of over 85 year olds (see Table 5.9).

Table 5.9 *Types of Alzheimer's disease*

Chromosomal (Down's Syndrome)	<1%
Familial	25%
Late onset	15% to 25%
Early onset	<2%
Sporadic	75%

Chromosomal (Down's syndrome) Alzheimer's disease

More than half of all individuals who have Down's Syndrome (trisomy 21) develop Alzheimer's disease after the age of 40 years. This is thought to be due to the over-expression of the APP gene on chromosome 21, which encodes for the amyloid precursor protein. Over-expression leads to the over-production of the beta amyloid in the brain.

Familial Alzheimer's disease

Approximately 25 per cent of all Alzheimer's cases are familial, in that two or more family members have also been affected. Most individuals who develop this condition do so in later life. Late onset is categorised as occurring over the age of 65 years, while early onset is at any age up to 65 years (see Table 5.10).

Table 5.10 *Familial Alzheimer's disease*

Late onset > 65 years	**Early onset < 65 years**
95% of all familial cases	5% of all familial cases
Polygenic multifactorial	Autosomal dominant

Late onset. Genetic studies suggest that late onset Alzheimer's disease is a multifactorial condition that involves numerous susceptibility genes. However, the identification of these genes remains elusive with the exception of the APOE gene that, to date, is the only confirmed genetic risk factor. The APOE gene is situated on chromosome 19q and its function is to encode for apolipoprotein E, which combines with lipids to form lipoproteins. These lipoproteins combine with blood cholesterol and transport the cholesterol to the liver for processing. There are many allelic variants of the APOE gene, one of which (allelic form e4) results in an increased risk for developing Alzheimer's disease. The e4 allele is associated with increasing number of amyloid plaques in the brain tissue.

Early onset. Three genes have been identified with early onset familial Alzheimer's disease. These genes display highly penetrant mutations that can be inherited in an autosomal fashion. The APP gene on chromosome 21q encodes for the amyloid beta protein, which is a major component of the brain deposits found in individuals with Alzheimer's disease. This altered APP gene accounts for 10 to 15 per cent of all early onset familial cases. The PSEN1 gene on chromosome 14q encodes for the protein presenilin-1. This protein is involved in the development and functioning of neurones. It also has an essential function in the processing of amyloid precursor proteins. More than 150 different types of mutations in the PSEN1 gene have been found in patients with Alzheimer's disease and defective gene copies account for 30 to 70 per cent of early onset familial cases. The PSEN2 gene on chromosome 1q encodes for the protein presenilin-2, which processes proteins that are needed for the transmission of chemical signals from the cell membrane into the cell nucleus. It is an essential protein that is needed for the activation of other genes. It also has a similar role to presenilin-1 in that both act as enzymes in cutting the amyloid precursor protein. Without this action, the amyloid precursor protein builds up in the brain, causing Alzheimer's disease. Alterations in the PSEN2 gene account for less than 5 per cent of all early onset cases.

3. Sporadic

Individuals with no family history of Alzheimer's disease account for approximately 75 per cent of all Alzheimer's disease cases. Sporadic forms of the condition display complex inheritance and are most probably a multifactorial condition. Genetic susceptibility factors remain elusive, with alterations to the APOE gene being the only confirmed genetic risk factor.

Late onset Alzheimer's disease results from a combination of genetic, ageing and environmental factors. No environmental agents such as head injury, viruses or toxins have been proven to be directly linked to the development of the condition to date.

Congenital disorders

The term 'congenital' means existing at birth and includes all birth defects regardless of cause. The frequency of congenital disorders is high but is reduced at birth due to the high level of miscarriages and abortions. About 14 per cent of all births have a recognised single defect, of which 25 per cent will have a recognised genetic cause. Extrinsic agents that are known to cause birth defects are known as **teratogens.** The susceptibility of the foetus to a teratogen is dependent upon its genotype. Neural tube defects and cleft lip/palate are just two examples of congenital multifactorial conditions.

1. Neural tube defects

Neural tube defects are birth defects that affect the brain and spinal cord. The two most common forms are anencephaly and spina bifida. Anencephaly occurs when the upper end of the neural tube fails to close during the first month of pregnancy. Infants born with this disorder do not have a developed brain and are usually stillborn or die shortly after delivery.

Spina bifida results in the lower part of the foetal spinal column failing to completely close during development. There is usually some accompanying nerve damage resulting in varying degrees of lower limb paralysis.

Neural tube defects are caused by a number of different genes and environmental factors. Alterations in the T locus on chromosome 6q have been associated with increased risk of spina bifida, and alterations in the MTHFR gene on chromosome 1 and the VANGL 1 gene on chromosome 1p have been found in individuals with familial neural tube defects.

Research has shown that women who have folic acid deficiency are at a higher risk of having a child with spina bifida. Taking folic acid supplements has been shown to significantly reduce, but not negate, the risk. Vitamin B12 deficiencies have also been linked to an increased risk of spina bifida, explained by the fact that Vitamin B12 forms an essential part of the folic acid pathway within the body. Research has also shown that some anticonvulsant medication, maternal diabetes, maternal obesity and hyperthermia in early development can also increase the risk of spina bifida.

2. Cleft lip/palate

This condition arises as a result of the failure of fusion of the frontal and maxillary processes during embryonic development in the first six to eight weeks of pregnancy. When these tissues fail to meet, a gap appears. This gap can appear in a single joining site or simultaneously in both lip and palate. It is a relatively common condition, affecting 1 in 1,000 Europeans, and has an empiric risk of 4 per cent for an affected sibling. The genetic component of cleft lip and palate has been identified in many syndromes, but understanding remains patchy in non-syndromic cases.

A few genes have been identified and linked to non-syndromic cases: the altered form of PVRL 1 gene on chromosome 11q and the altered MSX 1 gene on chromosome 4p are both linked with this condition. The incidence of cleft lip and palate is higher in boys (60 to 80 per cent of all cases), which could be linked to the MID 1 gene on the X chromosome (Xp). The variant forms of the IRF 6 gene on chromosome 1q have been identified with both non-syndromic and syndromic forms of cleft lip and palate (van der Woude syndrome).

Cleft lip and palate has a heritability rate of 76 per cent, which indicates that this condition is multifactorial. Studies have shown that in some syndromic cases, low levels of oxygen in early pregnancy can lead to the disorder. In Siderius X-linked syndrome, in which individuals are born with learning disabilities as well as a cleft lip and palate, the PHF 8 gene on chromosome Xp is affected by low oxygen levels. The catalystic activity of the PHF 8 enzyme requires adequate levels of oxygen to function. Links have also been made to maternal alcohol abuse, maternal anticonvulsive therapy, antihypertensive therapy, nitrate compounds and illegal drugs. Research is currently being focused on the role of folic acid in reducing the risks of this condition.

ACTIVITY 5.5

a. Define the following words.

i) Phenocopy.

ii) Pleiotropy.

iii) Penetrance.

iv) Heritability.

b. The incidence of obesity has tripled in the UK over the past 20 years and is still rising. Is this due to genetic or environmental factors? Explain your reasoning.

IDENTIFYING THE GENES IN MULTIFACTORIAL CONDITIONS

Most common disorders are polygenic and multifactorial. Identifying the genes that cause these conditions is highly problematic. One of the techniques used by genetic researchers is to examine the DNA of family members who have the same condition and compare their DNA sequences to the DNA of non-affected family members. Regions of DNA that show similarities between affected relatives, but are different to those of the non-affected relatives, are identified as 'susceptibility genes'. These studies are known as association studies, but they are limited in that they establish correlations but not causation. Compounding the problem of identifying the causative genes is that different genes can produce the same phenotype (genetic heterogeneity). This results in different 'susceptibility genes' being identified in different families. Hypertension, for example, can arise as a result of a variety of different causes (increased production of antidiuretic hormone, aldosterone, renin, erythropoietin or decreased elasticity within the arterial walls). Different causative effects will exist in different hypertensive individuals. Genetic heterogeneity in humans means that thousands of individual genotypes have to be examined before any correlations are found between the DNA sequence and the disease in question. Other barriers include traits that are phenocopies, where the cause is purely environmental and not genetic. Examining the DNA sequence in these traits is of no use or benefit.

A change in one gene can lead to variations of the same disorder or cause numerous different disorders. The LMNA gene, mentioned earlier in relation to cardiovascular disease, can also be linked to muscular dystrophy and to the ageing disease of Hutchinson–Gilford progeria syndrome. Identifying these genes can only be done by examining the genome of thousands of individuals.

Research is progressing, with large-scale projects being established throughout the world. In the UK, the UK Biobank project has been established to study the influence of genetics, environment and lifestyle on human disorders. The aim of the project is to study the genomes of half a million people.

Multifactorial conditions are common and account for the majority of human morbidity and mortality. However, rapid progress is being made towards the identification of both the genetic and environmental causes of disease through the use of large biobanks and continued research.

SUMMARY

- Monogenic traits are encoded for by a single gene. Monogenic conditions in humans are relatively rare.
- Working out the inheritance risk for monogenic traits can be done through the construction of a Punnet square. When considering more than one monogenic trait, the laws of addition and multiplication can be used to predict the probability of different outcomes.

- A polygenic trait is encoded for by a number of related genes. These genes tend to have an additive effect that gives rise to a continuous variation in the phenotype as, for example, with skin colour, height, weight and blood pressure.
- Both monogenic and polygenic traits can be influenced by external factors. These traits are termed multifactorial. Environmental influences can include oxygen levels, temperature, dietary factors, stress, activity levels, viruses and toxins.
- Multifactorial inheritance patterns are difficult to predict as they are complex disorders. The risk of inheriting a complex disorder is calculated by empiric risk and heritability. These two measurements apply to the population and not to the individual.
- The most common disorders of adult life are polygenic, multifactorial disorders. The 'susceptibility genes' of many of these disorders have been identified but knowledge regarding the environmental influences remains patchy.

FURTHER READING

Cox, T.C. (2004) 'Taking it to the max: the genetic and development mechanisms coordinating midfacial morphogenesis and dysmorphology'. *Clinical Genetics* 65(3): 163–76

Greene, N.D.E., Stanier, P. and Copp, A.J. (2009) 'Genetics of human neural tube defects'. *Human Molecular Genetics* 18(2): 113–29

Scourfield, J. and McGuffin, P. (1999) 'Familial risks and genetic counselling for common psychiatric disorders'. *Advances in Psychiatric Treatment* 5: 39–45

Shi, M., Wehby, G.L. and Murray, J.C. (2008) 'Review on genetic variants and maternal smoking in the etiology of oral clefts and other birth defects'. *Birth Defects Research,* 84(1): 16–29

For further details regarding any of the conditions mentioned or the actual genes involved see:

Online Mendelian Inheritance in Man, OMIM, McKusick-Nathans Institute of Genetic Medicine, Johns Hopkins University and National Centre for Biotechnology Information, National Library of Medicine **www.ncbi.nlm.nih.gov/omim**

Rare diseases. Genetic information and a list of relevant websites are produced by this American website **www.rarediseases.info.nih.gov**

An excellent virtual genetics site has been developed by Leicester University which has good simple explanations, good graphics and is easy to navigate: **www.le.ac.uk/ge/genie/vgec/index.html**

06

MUTATIONS

LEARNING OUTCOMES

The following topics are covered in this chapter:

- inherited and acquired mutations;
- effects of mutations;
- chromosomal mutations, numerical and structural abnormalities;
- gene mutations, point and frameshift mutations.

INTRODUCTION

A mutation is any change in DNA that can be reproduced. Normally the exact genetic information from both parents will be present in the offspring at fertilisation. However, sometimes errors occur that result in a mutation. A mutation gives dramatically different information from what is expected. These mutations can occur in the germ cells that can be passed on to the next generation, or in somatic cells that are not transferable to any offspring.

Inherited mutations

Germ line mutations occur in either the sperm cells or in the egg cells. The mutation then becomes inherited by the offspring. These mutations may therefore result in hereditary diseases.

Acquired mutations

Somatic mutations occur in the body cells and are not transmitted to the next generation, but are present in all descendants of that particular cell. This type of mutation can occur at any time during a person's lifetime. This type of mutation can be caused by some environmental factors such as radiation. A new mutation that has not been inherited is termed ***de novo***. Most cancers are caused by acquired mutations.

Polymorphisms

Changes that occur in more than one per cent of the population are termed **polymorphisms**. These are considered to be a normal variation in DNA. Polymorphisms are responsible for variations within populations as, for example, hair colour, eye colour and blood types. These polymorphisms have no negative effect on health but some may increase the risk of developing certain disorders.

ACTIVITY 6.1

Can you think of two ways in which an individual may acquire a mutation?

The effect of mutations

- **Less favourable**: a less favourable mutation can either be lethal, where the change is not compatible with life, or sub-lethal, which limits the individual's ability to grow to maturity.
- **Beneficial**: some mutations may have a positive effect. For example, a mutation has been recorded in individuals of European descent that gives resistance to the HIV virus in homozygotes.
- **Neutral**: the effects of these mutations do not influence the health of the individual. It is believed that the majority of mutations fall into this group.

Level of occurrence

A mutation can occur at any level of the genome. Whole chromosomes can be affected or just a single gene.

CHROMOSOMAL MUTATIONS

Any change in chromosome number or structure is called a **chromosomal abnormality**. These abnormalities usually occur when there is an error in cell division following meiosis or mitosis. There are many types of chromosome abnormalities but they can be classified into two groups: numerical and structural.

Numerical abnormalities

Cells normally contain 23 pairs of chromosomes and 46 individual chromosomes in total. A change in the number of chromosomes can occur during the formation of the sex cells (egg and sperm), in early embryonic development or in any cell after birth. Cells may contain an extra copy of a chromosome (**trisomy**) or may be missing a chromosome (**monosomy**).

Only a few numerical abnormalities found in gametes survive to birth. Survival seems to depend on which chromosomes are involved. Numerical abnormalities in chromosomes that support relatively low numbers of genes such as chromosome 13, 18 and 21 appear compatible with life. The X chromosome has a natural mechanism to adjust gene dosage and also becomes compatible with life. Examples of numerical abnormalities are shown in Table 6.1 and in Case studies 6.1, 6.2 and 6.3. Examples of sex chromosome abnormalities are shown in Table 6.2 and in Case studies 6.4, 6.5, 6.6 and 6.7.

Table 6.1 *Autosomal abnormalities*

Chromosome	**Syndrome**	**Incidence per 1000 births**	**Lifespan**
Trisomy 13	Patau's	0.2	1 month
Trisomy 18	Edwards'	0.3	1 year
Trisomy 21	Down's	1.5	50 years

CASE STUDY 6.1

Patau's syndrome

Clinical features

Babies born with Patau's syndrome may display a wide range of different features. The most common include neurological abnormalities arising from the failure of the development of paired cerebral hemispheres and microcephaly. This results in severe learning disabilities. Other common features are small, closely spaced eyes or absent eyes or even a centrally spaced eye. A large proportion of cases will have 'rocker bottom' feet and extra fingers and toes. Internal organs are also often affected, with renal dysplasia being common.

Genetics

Patau's syndrome is due to trisomy of chromosome 13.

Prognosis

Prognosis is poor. Life expectancy is generally up to 12 months, but 50 per cent die within the first month of life.

CASE STUDY 6.2

Edwards' syndrome

Clinical features

The most common feature of this disorder is severe developmental delay. Babies born with Edwards' syndrome tend to have a low birth weight, heart defects, a prominent occiput and severe learning difficulties. One of the classic signs of this disorder is the overlapping of the index finger and little finger over the middle two fingers and a clenched fist. Some babies are also born with spina bifida, but this seems to be a variable feature.

Genetics

Edwards' syndrome is due to trisomy of chromosome 18.

Prognosis

Prognosis is poor. Life expectancy is about 12 months of age with only 10 per cent surviving beyond the first year of life.

Increasing maternal age increases the risk of Edwards' syndrome.

CASE STUDY 6.3

Down's syndrome

Clinical features

A variety of medical problems are associated with Down's syndrome, which include congenital heart disease, hypothyroidism, digestive tract problems and learning difficulties. Clinical features include a flat facial profile, epicanthal folds, a single transverse palmar crease, open mouth with a protruding tongue and usually a space between the first and second toes. Individuals with Down's syndrome are at a higher than normal risk of developing infections, heart defects, epilepsy and early onset Alzheimer's disease.

Genetics

Down's syndrome is due to trisomy of chromosome 21 in over 95 per cent of cases. Less than 5 per cent of cases are due to familial causes.

Prognosis

Prognosis is relatively good. However, life expectancy is usually only about 50 years. This reduced life expectancy has been mainly attributed to cardiac problems.

Increasing maternal age increases the risk of Down's syndrome.

Table 6.2 *Sex chromosome abnormalities*

Chromosome	Gender	Syndrome	Incidence per 1000 births	Lifespan
Monosomy X	Female	Turner's	0.2	30 to 40 years
XXY	Male	Klinefelter's	1.0	Normal
XXX	Female	Triple X	1.0	Normal
XYY	Male	XYY	1.0	Normal

CASE STUDY 6.4

Turner's syndrome

Clinical features

The phenotype of individuals with Turner's syndrome is female, but sexual maturity is not achieved. Intelligence levels are normal although there may be specific patterns of learning difficulties. Individuals will normally be short in stature (under 5 feet in height), have a web between the neck and shoulders, a low posterior hairline and an increased 'carrying angle' at the elbow. Lymphatic abnormalities and congenital heart disease are often associated with Turner's syndrome. Growth hormone and oestrogen supplementation tends to be standardised care.

Genetics

Partial or complete loss of an X chromosome in females results in Turner's syndrome. This is due to the loss of the genes positioned on the short arm of the X chromosome (Xp missing).

Prognosis

Life expectancy is usually 30 to 40 years. The shortened life expectancy is mainly attributed to cardiac problems.

CASE STUDY 6.5

Klinefelter's syndrome

Clinical features

The phenotype of individuals with Klinefelter's syndrome is male, due to the presence of the Y chromosome. Clinical features include a tall stature with long limbs and large hands and feet. Individuals have male genitalia but rudimentary testes that do not produce sperm. Some feminine sexual characteristics are not completely suppressed as individuals may develop gynaecomastia (enlargement of breast tissue), slight widening of the hips and feminine hair distribution (although facial hair is masculine). There may be learning difficulties, curvature of the spine and osteoporosis.

Genetics

Individuals with Klinefelter's syndrome have one Y chromosome and multiples of the X chromosomes. The number of X chromosomes may vary from case to case as karyotypes can be XXY, XXXY, XXXXY, etc.

Prognosis

Life expectancy is normal.

CASE STUDY 6.6

Triple X syndrome

Clinical features

This syndrome is highly variable in its clinical presentation. More often than not, individuals do not display any clinical features. They develop normally both physically and mentally. However, there appears to be, in some cases, the underdevelopment of secondary sexual characteristics, sterility and learning disabilities.

Genetics

Triple X syndrome is due to the presence of an extra X chromosome in females. There have been rare cases reported of XXXX and XXXXX. Clinical symptoms of the underdevelopment of secondary sexual characteristics, sterility and learning disabilities appear more severe in cases with a higher X chromosome count.

Prognosis

Life expectancy is not affected.

CASE STUDY 6.7

XYY syndrome

Clinical features

Individuals with this condition are male and are usually tall in stature (over 6 feet in height). Fertility is not affected. Many individuals also display learning difficulties and some experience problems with motor co-ordination. Early claims by different texts that individuals displayed aggressive behaviour have not been substantiated.

Genetics

An extra Y chromosome in males. This is due to the inheritance of two Y chromosomes from the father.

Prognosis

Life expectancy is normal.

Mosaicism

If only a proportion of an individual's cells contain chromosomal abnormalities then the term **mosaicism** is used. This could be as a result of numerical abnormalities occurring in early embryonic development, resulting in the individual having some cells with numerical abnormalities and some normal cells.

ACTIVITY 6.2

In which chromosome can monosomy occur and still be compatible with life?

Structural abnormalities

A structural abnormality of a chromosome is when parts of an individual chromosome are either missing or duplicated. The abnormalities can range from the loss or duplication of a very small fragment up to the loss or duplication of a whole chromosomal arm.

Structural abnormalities can be classified as balanced or unbalanced. Balanced is when there is rearrangement of genetic material either within or between chromosomes. Unbalanced is where genetic material is either gained or lost.

Balanced

There are three types of balanced structural abnormalities:

1. **Translocation.** This is when a portion of one chromosome exchanges place with a portion from another chromosome. This is a reciprocal rearrangement of genetic material (see Figure 6.1).
2. **Inversion**. This is where genetic material becomes inverted. A segment of the chromosome breaks off, turns upside-down and then reattaches itself to the original chromosome. This occurs on a single chromosome (see Figure 6.2).
3. **Insertion.** This is where genetic material from one chromosome is inserted into another chromosome (see Figure 6.3). There is no reciprocal exchange here as in translocation.

Some types of hermaphrodism have arisen from this type of structural abnormality, where genetic material from the Y chromosome has been inserted into the X chromosome during meiosis.

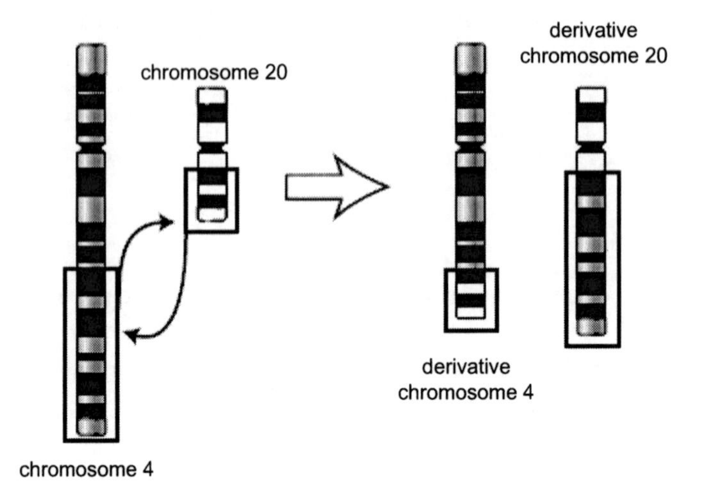

Figure 6.1 *Translocation*

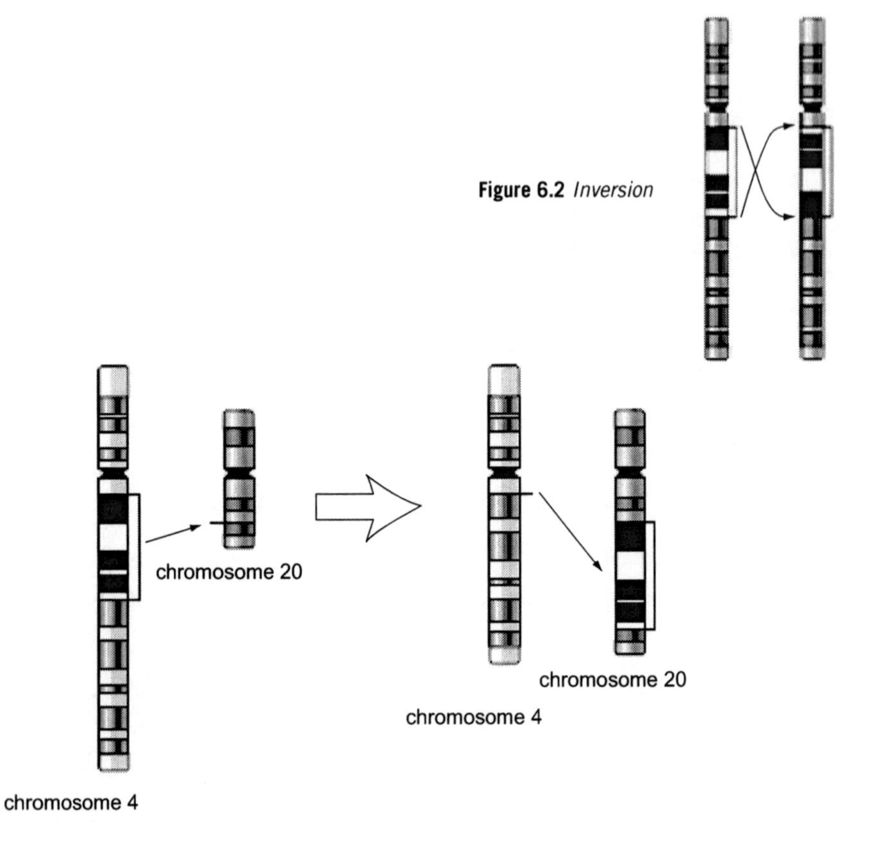

Figure 6.2 *Inversion*

Figure 6.3 *Insertion*

Examples of balanced structural abnormalities are shown in Case studies 6.8 and 6.9.

CASE STUDY 6.8

Chronic Myelogenous Leukaemia

Clinical features

Chronic myeloid leukaemia (CML) affects blood cells. Malignant leukaemic cells replace the normal blood cells in the bone marrow.Individuals with this condition show signs of anaemia, clotting disorders and are immunosuppressed.

Genetics

CML is caused, in 90 per cent of cases, by the translocation of chromosomal material. Translocation occurs between parts of the long arm of chromosome 9 (9q) and the long arm of chromosome 22 (22q). The reciprocal translocation between 9q and 22q leaves a characteristic modified version of chromosome 22, which has been called the 'Philadelphia chromosome' after the city where it was first discovered.

CASE STUDY 6.9

Familial Down's syndrome

Clinical features

Clinical features are the same as for Down's syndrome, which has been described previously as an autosomal numerical abnormality (see page 112).

Genetics

In 5 per cent of Down's syndrome cases the condition is familial.This is where there is translocation of genetic material between chromosome 14 and chromosome 21. The individual who develops this translocation has a 50 per cent chance of passing the affected chromosome 14 onto their offspring. If this occurs then the affected chromosome 14, which includes a large amount of chromosomal material from chromosome 21, affects the genome as there will be three copies of the genetic material of chromosome 21 present. The offspring in this case would have Down's syndrome. In individuals who possess this translocation, the risk of having a child with Down's syndrome is high.

Unbalanced

There are two types of unbalanced structural abnormalities.

1. Deletions

This is where a portion of the chromosome is lost (see Figure 6.4). This can be classified as terminal (near the tip of the chromosome) or interstitial (within the long or short arm). Examples of known conditions arriving from deletions are shown in Table 6.3 and Case studies 6.10 to 6.15.

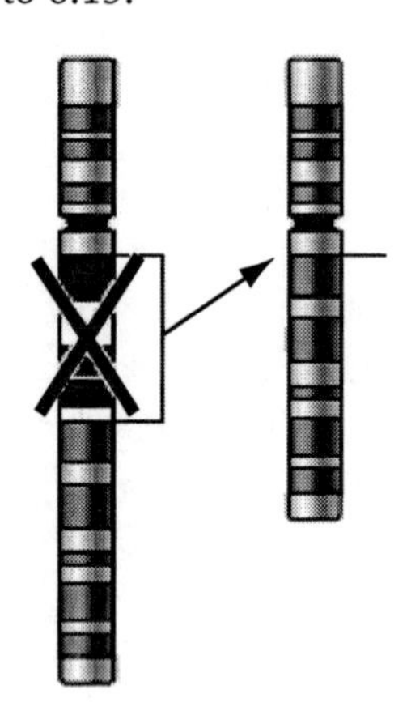

Figure 6.4 *Deletion*

Table 6.3 *Examples of known conditions arising from deletions*

Condition	Chromosome	Position
Cri-du-chat	5	Terminal
Wolf–Hirschhorn	4	Terminal
WAGR	11	Deletion within 11p
Prader–Willi/Angelman	15	Deletion within 15q
Di George	22	Deletion within 22q

CASE STUDY 6.10

Cri-du-chat syndrome

Clinical features

This syndrome is so named due to the presence of a high pitched cry, like a cat meowing, in newborns. The high pitched cry is due to a malformed larynx and is often used to diagnose the condition and to refer on for chromosomal analysis for confirmation. Other features include microcephaly, epicanthal folds, a round face, low-set ears and severe learning disabilities.

Genetics

Cri-du-chat syndrome is also known as chromosome 5p deletion, as the terminal part of the short arm of chromosome 5 is missing. The deletion of 5p can vary in size from an extremely small segment to the entire arm.

CASE STUDY 6.11

Wolf–Hirschhorn syndrome

Clinical features

This syndrome is characterised by severe growth retardation and profound cognitive impairment, microcephaly, closure defects such as cleft lip or palate and cardiac septal defects. Individuals with this condition have certain facial features such as a broad and prominent nose, epicanthal folds and low-set ears. Epilepsy is also common in this condition.

Genetics

Wolf–Hirschhorn syndrome is caused by the deletion of the terminal portion of the short arm of chromosome 4 (4p terminal deletion).

CASE STUDY 6.12

WAGR syndrome

Clinical features

WAGR syndrome stands for **W**ilms' tumour, **A**niridia, **G**enitourinary abnormalities and mental **R**etardation syndrome. Wilms' tumour is a malignant renal tumour that affects young children. Aniridia is where the individual is born without the presence of an iris in the eyes, although some individuals with WAGR may have iris displacement rather than aniridia. Genitourinary abnormalities can vary from patient to patient and demonstrate as ambiguous genitalia in some. There is a wide range of severity of learning disabilities in this syndrome. Not all aspects of WAGR are displayed in all individuals with this condition.

Genetics

WAGR syndrome is caused by the deletion of chromosomal material within the short arm of chromosome 11 (11p).

CASE STUDY 6.13

Angelman syndrome

Clinical features

This syndrome is characterised by a developmental delay in neurological abilities. Learning disabilities, poor limb coordination and distinctive behavioural patterns such as inappropriate excitement are present in most individuals with this syndrome. A high proportion of individuals also have heart problems, epilepsy, speech problems and minor facial dysmorphology.

Genetics

This syndrome is due to the loss of some genes from the long arm of chromosome 15 (15q), but only on chromosome 15 that has been inherited from the mother. The loss of the same genes inherited from the father gives rise to Prader–Willi syndrome, an entirely different condition.

CASE STUDY 6.14

Prader–Willi syndrome

Clinical features

This syndrome is characterised by diminishing foetal activity in the womb, severe hypotonia at birth and feeding difficulties up to the age of six months. Between the ages of one and six years, individuals with this condition develop hyperphagia (overeating), which leads to obesity. Other clinical features include short stature in adulthood and varying degrees of learning difficulties.

Genetics

Prader–Willi syndrome is due to the loss of genes from within the long arm of chromosome 15 (15q). A number of genes are lost from this region, not all of which have been identified as yet. The affected chromosome 15 is only inherited from the father and acts as an autosomal dominant inheritance. The loss of the same genes if inherited from the mother results in a completely different condition called Angelman syndrome. Prader–Willi syndrome occurs in individuals who do not express all of the genes on chromosome 15q.

CASE STUDY 6.15

Di George syndrome

Clinical features

A pattern of craniofacial, cardiovascular and immunological abnormalities exists in this syndrome. Individuals display facial dysmorphology, ventricular septal defects, underdevelopment of the parathyroid gland and an absence of the thymus gland. These features arise from the disturbed neural crest cell migration in the developing embryo. Many individuals have mild learning difficulties and up to 25 per cent develop psychiatric problems such as depression and schizophrenia.

Genetics

Di George syndrome is due to a deletion within the long arm of chromosome 22 (22q). About 30 genes are lost from this region in this syndrome.

2. Duplications

This is where a portion of a chromosome is duplicated, resulting in an exact copy of genetic material on the same chromosome (see Figure 6.5). A known condition caused by duplication of genetic material on chromosome 17 is Charcot–Marie–Tooth Disease Type A (Case study 6.16).

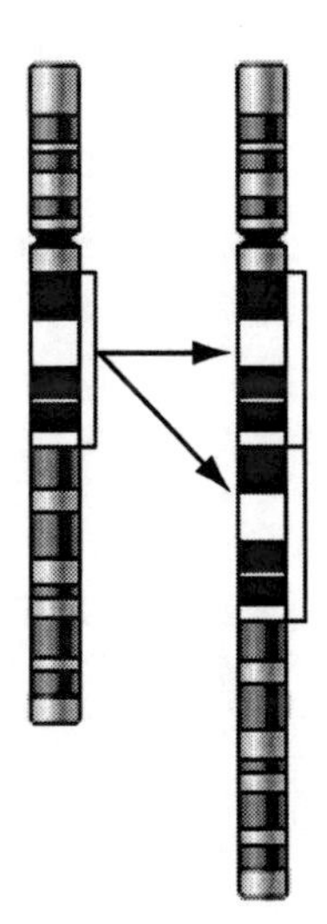

Figure 6.5 *Duplication*

The severity of chromosome structure abnormalities depends on whether individual chromosomes have been disrupted. Deletions are normally expected to be more clinically severe when compared with duplications.

CASE STUDY 6.16

Charcot–Marie–Tooth Type A syndrome

Clinical features

This is a neurological condition where there is progressive muscle weakness and severe neuropathy that mainly affects the lower legs and hands. Ankle weakness resulting in bilateral foot drop is common. This condition affects 1 in every 2,500. Life expectancy is not affected.

Genetics

The syndrome is caused by the duplication of the PMP22 gene (that codes for peripheral myelin) on the short arm of chromosome 17 (17p).This acts as an autosomal dominant disorder. Higher amounts of duplications are associated with increasingly severe symptoms.

ACTIVITY 6.3

With regard to the classification of structural abnormalities, cri-du-chat syndrome is an example of an unbalanced deletion. How would you classify familial Down's syndrome?

Factors that may increase the risk of chromosomal abnormalities

Maternal age

Females are born with all their eggs so, when a woman approaches her twentieth birthday, then so do her eggs. There is an identified link between maternal age and chromosomal abnormalities and this is thought to be due to the ageing process of the eggs. Males produce sperm throughout their lifetime. No links have been demonstrated between paternal age and chromosomal abnormalities.

Environment

There is, as yet, no conclusive evidence that environmental factors can result in chromosomal abnormalities.

ACTIVITY 6.4

How would you explain to a colleague the difference between a structural and a numerical mutation?

GENE MUTATIONS

A mutation at gene level can affect the base sequence of one allele. Variations in individuals can result from a mutation at gene level. Gene mutations can either be spontaneous or induced.

Spontaneous mutations

Random errors in DNA replication can occur spontaneously. These errors are common but most are detected and corrected by enzymes within the nucleus. The errors that do remain undetected have the potential to alter the phenotype in some way.

Induced mutations

An agent that can induce a mutation at gene level is called a mutagen. Known mutagens can be chemical, radiation or viral.

Mutations occurring during meiosis produce gametes that contain the abnormal base sequence in the DNA. This may result in an abnormal gene. The affected gene can be dominant or recessive and can occur on either the autosomes or on the sex chromosomes. Most mutations make the zygote incapable of completing normal development. Gene mutation (rather than chromosomal abnormalities) is probably the main cause of a high mortality rate among embryos.

If the mutation results in a dominant gene, and the gestation survival is not affected, then the individual's phenotype will display the results of the altered gene. If the mutated gene is recessive then the individual will become a carrier and will not show the effects in their own phenotype.

A recessive autosomal gene can lie undetected over many generations. The phenotype will be affected only when fertilisation occurs between two individuals who carry this gene and both pass on the recessive gene to the offspring. This individual, who is homozygous for the affected gene, will be the first to show the phenotypic effect of the mutated gene. An example of a dominant affected gene is Huntington's disease and a recessive affected gene is Cystic Fibrosis.

What happens at base level

As each set of three bases codes for one amino acid, a change in just one base can alter and disrupt protein synthesis. Mutation can result from either exchanging one base for another (**point mutation**) or by deleting or inserting an additional base (**frameshift mutation**).

Point mutations

A point mutation results in one base being substituted for another. This is similar to a grammatical error in written language when a word is spelt incorrectly. The following example of a point mutation uses the analogy of language to demonstrate a point mutation. The original, non-mutated information might be:

WHY DID DAN EAT COW PIE

With a point mutation this might become:

WHY DID MAN EAT COW PIE

Or even:

WHY DID DAN EAT SOW PIE

A change in just one letter alters the sense of the message. Clinical examples of point mutations are Sickle Cell Anaemia and Thalassaemia (Case study 6.17).

CASE STUDY 6.17

HAEMOGLOBINOPATHIES

Normal haemoglobin is composed of two alpha polypeptide chains and two beta polypeptide chains. The function of these four chains is to carry oxygen molecules around the body in red blood cells. Each chain is able to carry one molecule of oxygen. Haemoglobinopathies is a term that relates to a wide range of autosomal recessive disorders in which the structure of either the alpha chain or beta chain is altered or missing.

Sickle Cell Anaemia

Clinical features

This condition is caused by abnormal beta chain synthesis. Clinical symptoms include night sweats, coughs, limb and joint pains, poor appetite and fatigue. In individuals who have two recessive genes for this condition (homozygous recessive) the abnormal haemoglobin aggregates, causing the red blood cells to become sickle shaped. These mis-shaped red blood cells clog up the small blood vessels. Individuals who are carriers of this condition have normal biconcave shaped red blood cells but, in times of stress when oxygen tension becomes low, these normal cells also become sickle shaped.

Genetics

Sickle Cell Anaemia is caused by a point mutation on the long arm of chromosome 11 (11p). The base T is substituted for the base A. This results in the alteration of the protein produced as the amino acid glutamine is replaced by the amino acid valine.

This substitution does have the advantage of giving the individual affected a resistance to the malarial parasite. A high proportion of Sickle Cell heterozygotes can be found in geographical areas where malaria is prevalent. Death as a result of malaria only occurs in individuals who do not possess the sickle cell gene.

Thalassaemia

Clinical features

Thalassaemia is a functional deficiency of the haemoglobin peptide chains. Alpha Thalassaemia is a condition where the alpha polypeptide haemoglobin chain is either missing or non-functioning. Beta Thalassaemia is as a result of an affected or missing beta chain. In the absence of either one of the alpha or beta chains, the corresponding partner chain is produced at an excessive rate and eventually damages the membrane of the red blood cell, leading to the death of the cell.

Genetics

Alpha Thalassaemia is due to a point mutation on chromosome 16.
Beta Thalassaemia is due to a point mutation on chromosome 11.

Frameshift mutations

A single base may be added or deleted. This results in the whole sequence of bases following the inserted/deleted point being out of alignment for the correct coding of amino acids.

Insertion

If an additional letter is inserted into the earlier example of WHY DID DAN EAT COW PIE, this will result in a totally different message:

WHY DID DAA NEA TCO WPI E

Tay–Sachs disease (Case study 6.18) occurs due to a frameshift mutation resulting from the insertion of four bases.

CASE STUDY 6.18

Tay–Sachs disease

Clinical features

Tay–Sachs disease is an inborn error of metabolism. Individuals with this condition lack the enzyme hexosaminidase A, which is needed for lysosomal storage. This leads to a progressive deterioration of neurological function. Infants present with hypotonia and fail to develop motor skills. Epilepsy is also common. Tay–Sachs disease is lethal by the age of three or four years.

Genetics

Tay–Sachs disease is caused by a four-base insertion on chromosome 15.This results in a frameshift mutation, which results in a premature termination of the protein so that no functioning hexosaminidase A enzyme is produced.

Deletion

By deleting a letter, a similar effect of altering the message occurs:

WHY DID DNE ATC OWP IE

The earlier in the sequence that this occurs, the more altered the protein becomes.

Codon repeats

Another type of mutation through insertion at base level is where codons are repeated. This is a form of 'stuttering' in the base language of DNA. For example:

WHY DID DAN EAT COW PIE

might become

WHY DID DAN DAN DAN DAN EAT COW PIE

Myotonic Dystrophy, Huntington's disease and Fragile X syndrome are all examples of codon repeats (Case studies 6.19, 6.20 and 6.21).

Codon repeat disorders can expand from one generation to the next and are therefore called **dynamic mutations**. The increasing number of repeats in each generation appears to correspond with the increase in severity of the condition and the earlier onset of the symptoms.

CASE STUDY 6.19

Myotonic Dystrophy

Clinical features

This is a highly variable phenotype and age of onset varies considerably. In adults there is progressive muscle weakness, with some individuals displaying cardiac problems arising from cardiac conduction abnormalities and cataracts. Severe forms of Myotonic Dystrophy can be seen in infants who display hypotonia, breathing difficulties and learning disabilities.

Genetics

This condition arises as a result of codon repeats of CTG on the long arm of chromosome 19 (19q). It is inherited as an autosomal dominant condition.The amount of codon repeats can increase from one generation to the next, which can increase the severity of the condition and alter the age of onset. Expansions of codon repeats for this condition occur in the mother rather than the father.

CASE STUDY 6.20

Huntington's disease

Clinical features

This is a neurological disorder that is characterised by spasmodic, involuntary movements of the limbs and facial muscles, often accompanied by hypotonia. Dementia is also a clinical feature of this condition. Age of onset is usually between 25 and 50 years of age, although this is very variable.

Genetics

Huntington's disease is caused by codon repeats of CAG within the Huntingtin protein. The gene that codes for the Huntingtin protein is situated on the short arm of chromosome 4 (4p). Between 6 and 35 codon repeats within this gene are normal and show no adverse effects, but the disease presents itself in individuals who have more than 37 codon repeats. Affected individuals have anywhere between 37 and 121 codon repeats. The amount of repeats affects severity of symptoms and age of onset. Expansion of repeats occurs in the father rather than the mother. Inheritance is autosomal dominant.

CASE STUDY 6.21

Fragile X syndrome

Clinical features

Learning disabilities are present in all male individuals who have this syndrome. Males also display behavioural problems. Only half of females who have Fragile X have learning disabilities (the second normal gene appears to ameliorate the phenotype). Adult males tend to have long, narrow faces, prominent ears and macroorchidism (enlarged testes).

Genetics

This condition is due to excessive codon repeats of CGC on the long arm of the X chromosome. No adverse effect is seen up to 60 copies of this codon. However, 60 to 200 repeats is premutation (can easily expand in future generations). Over 200 repeats have the effect of the failure to produce the protein FMR1, which causes the learning disabilities associated with this syndrome. In individuals who have between 60 and 200 repeats (premutation carriers),premature ovarian failure is common in women and tremor and ataxia are common in elderly males. Expansion of the codon repeats for this condition occurs in the mother rather than the father.

The classification of gene mutations

Mutations that change the DNA base sequence can be classified as:

- **Silent mutations**: the change in base results in the same amino acid being coded for.
- **Missense mutations**: the change results in the coding for a different amino acid.
- **Nonsense mutations**: this does not code for an amino acid and therefore terminates the protein structure.

ACTIVITY 6.5

a. If a codon mutated from AGC to AGA, would you classify this as a point mutation or a frameshift mutation?

b. Would the mutation from AGC to AGA be classified as a silent mutation, a missense mutation or a nonsense mutation?

Hint: you will need to refer to the codon-amino acid table on page 15 in Chapter 1 for the answer to (b).

Results of DNA mutations

The outcome of the mutation is dependent on its position within the genome. It could result in any of the following:

- No effect – for example, a change in base still codes for the same amino acid.
- Loss of function – for example, the protein is destroyed.
- Gain of function – for example, a new protein is created.
- Produces a structurally different protein but has the same function as the original protein.

SUMMARY

- Inherited mutations occur in either sperm or egg cells and are inherited from one generation to the next. Acquired mutations occur in the body cells and are not inherited.
- Changes that occur in more than 1 per cent of the population are termed polymorphisms. These account for the variation between individuals.
- Mutations can have a variety of effects, ranging from being lethal to being beneficial.
- Chromosomal mutations can either be the addition or deletion of a whole chromosome, or result in a structural abnormality of an individual chromosome.
- Gene mutations affect the base sequence of one allele. These mutations can be spontaneous or induced.

FURTHER READING

www.biology-online.org/2/8_mutations.htm
This website provides a good simple tutorial on genetic mutations which is part of the biology online book.

http://ghr.nlm.nih.gov
This is the Genetics Home Reference site, which is hosted by the United States National Library of Medicine. The website has a wide range of information available. *Your Guide to Understanding Genetic Conditions* gives a clear and detailed account on mutations, while the 'Genes and Diseases' section provides chromosome map links where you can click on a chromosome and identify the genes responsible for different conditions.

http://health.nih.gov/topic/GeneticDisorders
This website has a comprehensive list and links to common genetic conditions that you might encounter in the clinical area.

www.ncbi.nlm.nih.gov/omim
The Online Mendelian Inheritance in Man website is hosted by Johns Hopkins University in America. It is an excellent site for further detailed information on specific genetic conditions as it acts as a library for all relevant research publications on genetic disorders. It also has some excellent links to other good genetic sites.

07

PEDIGREE ANALYSIS

LEARNING OUTCOMES

The following topics are covered in this chapter:

- the use of pedigree charts;
- common patterns of inheritance;
- limitations of pedigree analysis.

INTRODUCTION

Pedigree analysis is the study of the inheritance of genes in humans. It is a systematic analysis of the familial history of an individual. A clear pedigree diagram is drawn where inheritance patterns can be determined.

Who's who?

Information is usually gathered from an individual who is requesting genetic advice. This individual is known as a **consultand.** The affected individual is usually termed the **proband.** If it is the affected individual who is requesting advice then they are both the consultand and the proband.

Information is gathered and documented in the form of a family tree. Recognised pedigree symbols are used so that family relationships are represented in a diagrammatical form (see Figure 7.1). These diagrams make it easier to recognise relationships within families.

Pedigrees are regularly used to determine the mode of inheritance (**dominant, recessive, sex-linked,** etc.) of genetic disorders as well as to discover the **probability** of passing the gene in question to the next generation.

The aims of pedigree analysis are to determine the mode of inheritance and to determine the probability of an affected offspring.

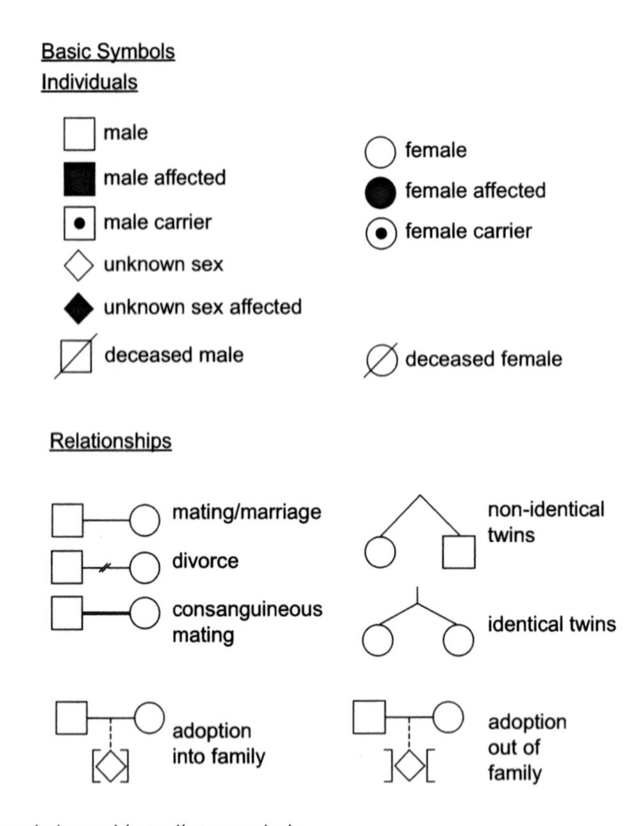

Figure 7. 1 *Basic symbols used in pedigree analysis*

The rules of pedigree analysis

Males are represented by squares, females by circles and individuals of unknown sex by diamonds. A horizontal line between squares and circles indicates marriage or mating between the male and female. Vertical lines extending downwards from a couple represent their biological offspring (see Figure 7.2). Subsequent generations are written underneath the parental generations with the oldest generation at the top of the pedigree chart.

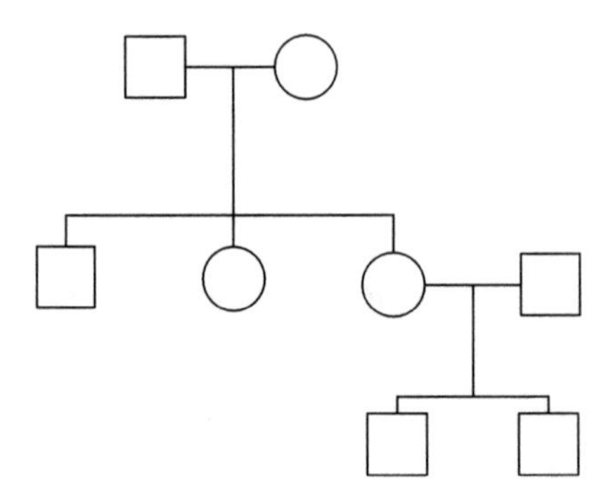

Figure 7.2 *The basic framework of a pedigree chart*

In Figure 7.2 it can be seen that the grandparents had three children, one son and two daughters. The birth order is written from left to right so, in this example, the eldest was male and he has two younger sisters. The youngest child (a daughter) had two children, both boys.

The purpose of a pedigree chart is to analyse the pattern of inheritance. Affected individuals are represented by a solid block of colour (see Figure 7.3).

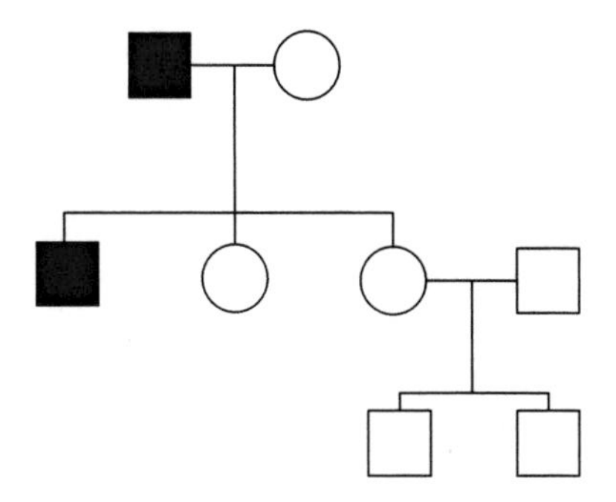

Figure 7.3 *Pedigree chart showing affected individuals*

It can be seen from this pedigree that the grandfather was affected and his son inherited this trait and was also affected. None of the daughters or the grandchildren is affected.

Generations in a pedigree chart are numbered from the top (eldest generation) downwards (youngest generation) by using Roman numerals. Individuals within the same generation are identified by using Arabic numerals from left to right (see Figure 7.4).

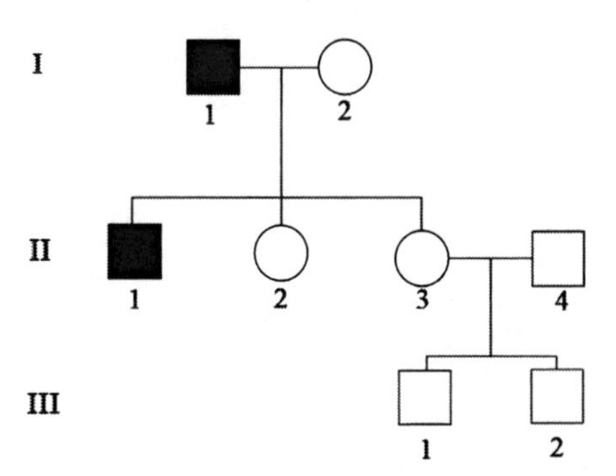

Figure 7.4 *Numbering a pedigree chart*

The affected individuals from this family can now be identified as I 1 and II 1.

Practical tips for drawing a pedigree chart:

1. Start at the centre of the page.
2. Collect details on at least three generations.
3. Enquire about:
   - stillbirths
   - miscarriages;
   - terminations;
   - neonatal deaths;
   - multiple partnerships;
   - consanguinity of partners;
   - deceased relatives;
   - adoptions.
4. Include details regarding:
   - ethnicity;
   - names;
   - dates of births;
   - medical diagnosis.

ACTIVITY 7.1

Draw a pedigree chart using the following information.

James and Louise are about to have their second child. Their first child John has already been diagnosed with a specific genetic condition. Neither James nor Louise is affected with this condition. The only other member of the family that has been identified with this condition is James' maternal grandfather. None of Louise's parents or any of her late grandparents has been identified with this particular condition. Both James and Louise have siblings. James has an older sister called Grace and Louise has a younger sister called Charlotte, none of whom has children.

Hint: you should have a pedigree chart showing four generations.

ACTIVITY 7.2

From the pedigree chart that you have constructed in Activity 7.1, what number within the pedigree is James?

Hint: Remember that you need to use both Roman and Arabic numerals.

Check your answers to Activity 7.1 and Activity 7.2 against the answers given at the end of the book.

MODES OF INHERITANCE

Most genes are inherited in a Mendelian fashion. When a variant form of a gene that causes a disorder is present, the inheritance pattern can be deduced from a pedigree chart. Certain rules exist for different types of inheritance.

Autosomal dominant inheritance

- Every individual will have an affected parent.
- The disorder does not 'skip' generations.
- Two affected individuals may have an unaffected child (if both parents are heterozygous) (see Figure 7.5).

In the pedigree chart in Figure 7.5, I 2 must have been heterozygous for the dominant trait as not all offspring have been affected (the recessive gene has been passed from I 2 to II 4).

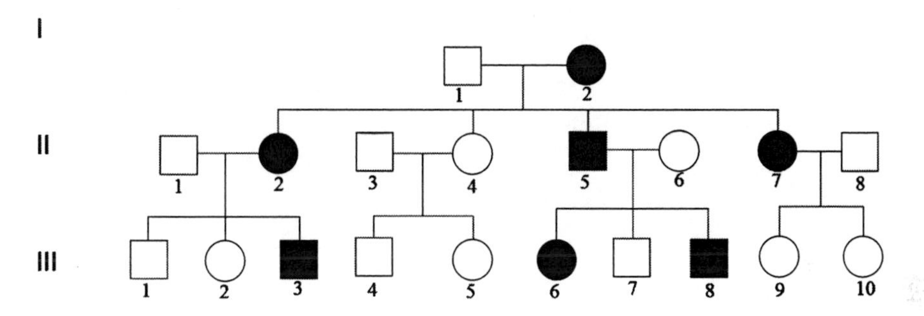

Figure 7.5 *An example of autosomal dominance within a family*

Examples of autosomal dominant disorders

- Huntington's disease (see page 127 in Chapter 6 for clinical features and gene details).
- Neurofibromatosis (see page 169 in Chapter 9 for clinical features and gene details).
- Myotonic Dystrophy (see page 126 in Chapter 6 for clinical features and gene details).
- Achondroplasia (Case Study 7.1).
- Marfan Syndrome (Case Study 7.2).

CASE STUDY 7.1

Achondroplasia

Clinical features

Achondroplasia is the most common form of dwarfism. Individuals born with this condition have a short stature due to the shortening of the limbs. Reduced muscle tone (hypotonia) is often present at birth, which leads to an increased risk of the cessation of breathing. Hyperextension of joints is common, especially of the knees, but the extension and rotation of the elbows is very limited. Exaggerated lordosis in the lumbar section of the spine develops as soon as a child starts walking. Individuals usually have a large head and the bridge of the nose is usually depressed. Intelligence is usually normal.

Genetics

This is an autosomal dominant disorder.The identified mutated gene for achondroplasia is the FGFR3 gene, which is situated on the short arm of chromosome 4 (4p). Affected individuals are heterozygous for this condition, as two mutated FGFR3 genes are lethal. Over 80 per cent of achondroplasia cases are as a result of a new mutation. This is where there is no family history of the condition and it is caused by a *de novo* mutation (a mutation that is not inherited). The *de novo* mutations for achondroplasia are associated with the paternal copy of chromosome 4.

CASE STUDY 7.2

Marfan syndrome

Clinical features

This condition has variable features in different individuals. It is a connective tissue disorder that can affect the skeletal tissue, lungs, heart and eyes. Individuals with this condition tend to be quite tall with disproportional long limbs and fingers. Scoliosis and pectus excavatum (funnel chest) may also be present in some cases. Cardiovascular problems are common, especially mitral valve prolapses and weakening of the aortic arch, which can lead to aortic aneurysms and eventually death. A variety of eye disorders have been associated with Marfan syndrome, ranging from myopia to lens dislocation.

Genetics

Marfan syndrome is an autosomal dominant condition arising from a mutated form of the FBN1 gene. This gene is located on the long arm of chromosome 15 (15q). Several forms of mutation have been discovered within the FBN1 gene (see Online Mendelian Inheritance in Man for more details, reference: omim #154700). Most individuals with this syndrome will have inherited the mutated FBN1 gene from a parent. However, 25 per cent of cases occur through *de novo* mutations.

Autosomal recessive inheritance

- All individuals who are affected may have parents who are not affected.
- The disorder can 'skip' generations.
- All the children of two affected parents are affected (see Figure 7.6).

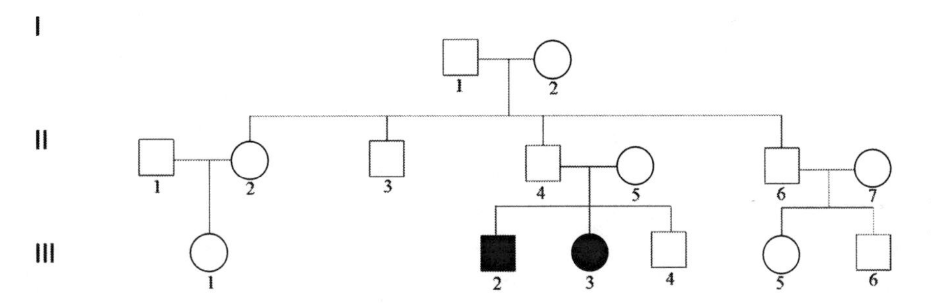

Figure 7.6 *An example of autosomal recessive inheritance in a family*

- Both III 2 and III 3 are homozygous for this trait.
- II 4 and II 5 can be identified as heterozygous and are therefore both carriers of the gene in question.
- Either I 1 or I 2 is also a carrier of this gene.

Examples of Autosomal Recessive Inheritance

- Cystic Fibrosis (Case study 7.3).
- Phenylketonuria (Case study 7.4).

CASE STUDY 7.3

Cystic Fibrosis

Clinical features

Cystic Fibrosis is a disorder that involves the accumulation of thick mucus in both the lungs and the digestive tract. This mucus accumulation is the result of defective chloride ion transport through cell membranes.The most common signs and symptoms are coughing and chronic breathing difficulties due to the accumulation of thick mucus in the lungs. This makes the individual susceptible to infections, which can lead to lung tissue damage over time. The digestive system is also affected, in that the thick mucus will prevent the pancreatic digestive enzymes from being released from the pancreas and reaching the small intestine. Weight gain is slow in children for this reason. Males with this disorder are nearly always sterile due to the absence of the *vas deferens* (the tube that carries the sperm). Cystic Fibrosis is life-limiting due to the scarring of the lung tissue from repeated infections.

Genetics

Cystic Fibrosis is due to a mutation in the CFTR gene. This gene is situated on the long arm of chromosome 7 (7q). The CFTR gene codes for chloride ion channels on the surface of cells in many organs. Mutations involve a deletion within the DNA resulting in the loss of the amino acid phenylalanine from the coded protein that forms the chloride ion channel. Around 70 per cent of Caucasians with Cystic Fibrosis carry this type of mutation. The remaining types of mutations affect other proteins that are essential in the transport of the CFTR protein out of the cell. Different types of mutations are found among different ethnic groups. Deletion in the DNA coding leading to the prevention of the CFTR protein from reaching its site of action is the most common cause of Cystic Fibrosis among all ethnic groups.

CASE STUDY 7.4

Phenylketonuria (PKU)

Clinical features

PKU is an inborn error in metabolism. Normally, the amino acid phenylalanine, gained from diet, is converted into the amino acid tyrosine. This metabolic action takes place in the liver by the action of the enzyme phenylalanine hydroxylase. This enzyme is defective in individuals with PKU. This results in increased levels of phenylalanine in the body and insufficient amounts of tyrosine. The amino acid phenylalanine is able to cross the blood–brain barrier and increased levels in babies can result in learning disabilities, and in adults can result in reduced cognitive functioning. High levels during pregnancy are teratogenic (embryonic development is affected) and may lead to microcephaly and congenital heart disease. The effects of insufficient levels of tyrosine are seen as hypopigmentation of both skin and hair.

Genetics

The phenylalanine hydroxylase enzyme is coded for by a single gene, situated on the long arm of chromosome 12 (12q). The condition is autosomal recessive, so an individual needs to inherit two genes that are both in a mutated form to have PKU. However, in heterozygotes (one mutated form only), the enzyme is only active at about a 30 per cent level. The frequency of the mutated forms of this gene varies considerably from one country to another. For example, it is extremely rare among the Japanese (1 in 119,000) but much more common among the Scots (1 in 5,000).

Dominant versus recessive

- If two affected people have unaffected offspring: **dominant gene**.
- If two unaffected people have affected offspring: **recessive gene**.

ACTIVITY 7.3

Draw a pedigree chart from the following information.

George and Samantha have three sons. Their middle son Josh has the same genetic condition as his mother Samantha. Neither of Josh's siblings is affected. Josh's maternal grandfather has also been identified with the same genetic condition. Neither George nor his parents has this condition.

Does your pedigree chart demonstrate the pattern for autosomal dominant inheritance or autosomal recessive inheritance?
Hint: relate your findings to the dominant and recessive rules. Check your work against the answer given at the end of the book.

Y-Linked Inheritance

- Traits on the Y chromosome are only found in males.
- The father's traits are passed to all sons, but never to daughters.
- Dominance is irrelevant as there is only one copy of the Y chromosome.

Figure 7.7 shows a typical pedigree chart of a Y-linked inheritance where all the males are affected.

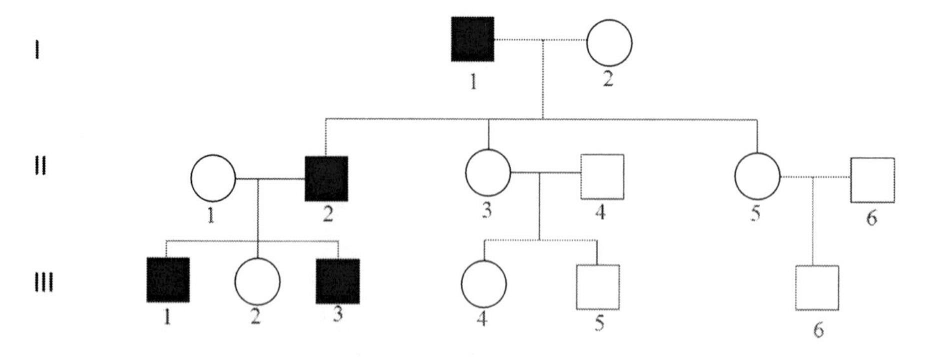

Figure 7.7 *An example of Y-linked inheritance in a family*

> **Examples of Y-linked inheritance:**
>
> There are only a few functional genes present on the Y chromosome, but one condition that can arise from deletions on the Y chromosome is Azoospermia (this condition results in lowered fertility, so the condition is rarely passed on) (Case study 7.5).

CASE STUDY 7.5

Azoospermia

Clinical features

Azoospermia is the absence of any living sperm in the semen. Males with this condition are therefore infertile, although some men do have clinically retrievable active sperm in their testes.

Genetics

This condition results from the deletion of DNA bases from the distal region of the long arm of the Y chromosome (Yq). Due to the clinical effects of this deletion, azoospermia is not usually inherited and most cases are *de novo* mutations. If active sperm is clinically retrieved from the testes and used for fertilisation, azoospermia is transmitted to all male offspring.

Sex-linked dominant inheritance

- These are dominant genes located on the X chromosome. Mothers pass one of their X chromosomes to both sons and daughters. Fathers only pass their X chromosome to their daughters.
- As females have two X chromosomes and males only one, males will express all X chromosome genes whether dominant or recessive.

- If a father is affected by a dominant gene, none of his sons will be affected, but all his daughters will be affected (see Figure 7.8).

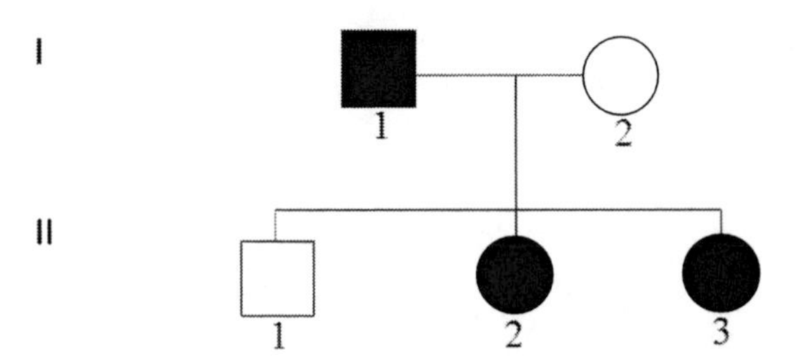

Figure 7.8 *An example of a pedigree chart for a sex-linked dominant disorder inherited from the father*

- If a mother is heterozygous for the dominant gene, and the father unaffected, their offspring have a 50 per cent chance of inheriting this gene (irrespective of their gender) (see Figure 7.9).

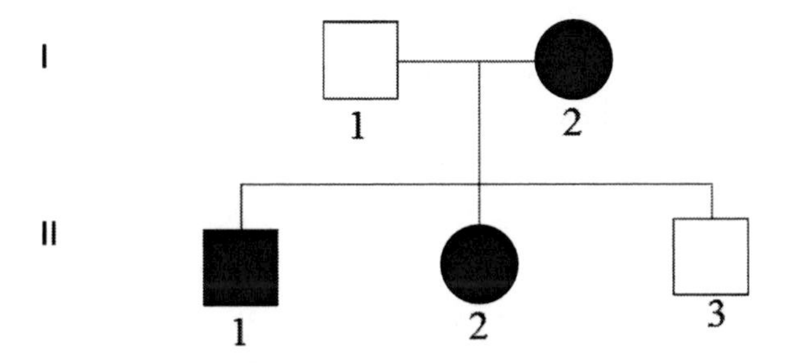

Figure 7.9 *An example of a pedigree chart for a sex-linked dominant disorder inherited from the mother*

Example of sex-linked dominant inheritance

Hypophosphataemia (Case study 7.6)

CASE STUDY 7.6

Hypophosphataemia

Clinical features

Hypophosphataemia is a form of rickets that arises from the inadequate reabsorption of phosphate in the proximal convoluted tubules of the kidney nephrons. Phosphate is needed for normal bone strength and growth. Individuals with this condition excrete high amounts of phosphate in their urine and have very low levels of phosphate in their blood. The effects of this condition, as with nearly all sex-linked disorders, are variable in intensity. Individuals can be short in stature with bowing of the lower limbs, suffer from bone pain and have bony outgrowths that can restrict joint movement. In severe cases the skull bones can fuse together too early, which can result in seizures. Females with hypophosphataemia tend to have less severe bone disease than males.

Genetics

This disorder arises due to a mutated form of the PHEX gene (Phosphate Regulating Endopeptidase X), which is situated on the X chromosome. The PHEX gene is a dominant gene, although the presence of a non-mutated form in females tends to lessen the severity of the effects of this disorder.

Sex-linked recessive inheritance

- Males get their X chromosome from their mother, while females inherit from both parents.
- Females express only if they get a recessive gene from both parents.
- Males will always express the recessive gene.
- Recessive disorders on the X chromosome are expressed much more in males than females (see Figure 7.10).

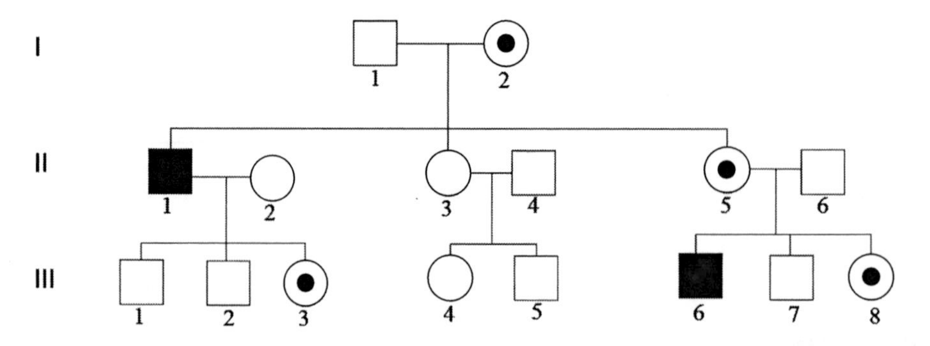

Figure 7.10 *An example of a pedigree chart that demonstrates an X-linked recessive disorder. Note how the carrier females are represented as a circle with a dot in the middle*

Examples of sex-linked recessive inheritance

- Haemophilia (factor VIII/factor IX) (Case study 7.7).
- Duchenne Muscular Dystrophy (Case study 7.8).

CASE STUDY 7.7

Haemophilia

Clinical features

Haemophilia is a common blood clotting disorder that is either due to a deficiency of clotting factor VIII (Haemophilia A) or clotting factor IX (Haemophilia B, also known as Christmas disease). Excess bleeding occurs as the clotting cascade is interrupted, resulting in insufficient levels of thrombin being produced. The severity of the condition depends on the level of factor activity present. In severe cases there is less than 1 per cent activity, resulting in bleeding into the joints and soft tissues. Individuals who have less than 35 per cent of clotting factor activity suffer from prolonged bleeding before clotting can occur.

Genetics

Haemophilia A results from a mutation within the Factor 8 (F8) gene and Haemophilia B results from a mutation within the Factor 9 (F9) gene. Both of these genes are located on the long arm of the X chromosome (Xq). Mutations in these genes normally arise from the inversion of genetic material (although deletions, insertions and point mutations have been noted in some cases).

CASE STUDY 7.8

Duchenne Muscular Dystrophy

Clinical features

Duchenne Muscular Dystrophy develops as a result of a dystrophin protein. The dystrophin protein is needed to stabilise muscle fibres. Without this protein muscle fibres weaken and die over time. The main muscles affected by this disorder are the skeletal muscles and the heart muscles. The effects can be seen in early childhood, with muscle weakening resulting in delayed motor skills such as sitting, standing and walking. The disease progresses rapidly, with most teenagers being wheelchair bound and suffering from cardiomyopathy. The heart muscles enlarge and weaken and limit life.

Genetics

Duchenne Muscular Dystrophy arises as a result of a mutation in the dystrophin gene, which is situated on the short arm of the X chromosome. The dystrophin gene is the longest known gene and hundreds of different mutations have been discovered in this gene (see Online Mendelian Inheritance in Man for more details, reference: omim #310200). In very rare instances where females have Muscular Dystrophy they either carry two mutated dystrophin genes or have an inactivated X chromosome. Females who are carriers are at an increased risk of developing muscle weakness and cardiomyopathy. Muscular Dystrophy is inherited in an X-linked manner in two thirds of cases, with the remaining third being due to *de novo* mutations.

Mitochondrial inheritance

The mitochondrion has its own genome. These genes are only inherited from the maternal line. All the mitochondria in the sperm are packed tightly in the tail. When fertilisation occurs, only the head of the sperm penetrates the ovum, leaving the tail outside. No mitochondrial genes are inherited from the father.

Dominant and recessive genes are irrelevant here as a full set of mitochondrial DNA is inherited from the maternal line (see Figure 7.11).

ACTIVITY 7.4

Elizabeth is colour blind. Colour blindness is an example of a sex-linked recessive gene. The gene for colour blindness is carried on the X chromosome. As it is a recessive gene, Elizabeth carries two copies of this gene (she is homozygous recessive for colour blindness). Elizabeth's husband has full colour vision. They have three sons. Elizabeth's mother is not colour blind but her father is.

Draw a pedigree chart by using the available information and identify if any of the three sons are also colour blind. Check if your answers are correct by comparing your answers with those given at the end of the book.

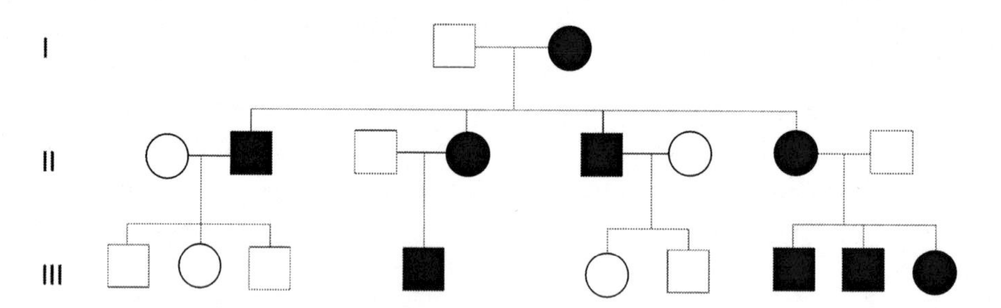

Figure 7.11 *An example of mitochondrial inheritance*

Examples of mitochondrial inheritance

- Kearns–Sayre syndrome (Case study 7.9).
- Leber's Hereditary Optic Neuropathy (Case study 7.10).

CASE STUDY 7.9

Kearns–Sayre syndrome

Clinical features

This is a neuromuscular disorder that begins in childhood. In most cases there is limited eye movement and the drooping of eyelids. Cardiac problems are common, especially electrical conductivity problems that lead to heart block. Hearing loss can occur and individuals are at an increased risk of developing diabetes.

Genetics

Kearns–Sayre syndrome arises as a result of DNA base deletions from various segments of the mitochondrial genome. Inheritance is only through the maternal line, as with all mitochondrial mutations.

CASE STUDY 7.10

Leber's Hereditary Optic Neuropathy

Clinical features

Individuals with this disorder usually have normal vision in childhood, but in young adulthood vision becomes blurred and eventually leads to blindness. This is due to the degeneration of the inner cellular layer of the retina and the retinal ganglion layer. Midlife symptoms of tremors and Multiple Sclerosis-type symptoms are common.

Genetics

This syndrome is due to mutations within mitochondrial DNA. Some individuals who carry these mutations do not go on to develop the disorder as there may be enough healthy mitochondrial DNA present. There is no way at present of predicting which carriers will go on to develop the disorder.

PROBLEMS WITH PEDIGREE ANALYSIS

There are some difficulties with establishing inheritance from pedigrees:

1. The same gene might have very different effects in different individuals. For example, Marfan syndrome (see page 136) might present as long limbs in one individual, bad eyesight in another and heart problems in yet another. All may have the same gene, but inheritance is not clear.
2. Hidden inbreeding. For very rare recessive disorders it often turns out that parents are related to each other, i.e. both are heterozygous because of common descent. The rarer the

condition, the higher the proportion of consanguineous marriages. For example, Cystic Fibrosis (see page 138) shows little difference in frequency between consanguineous and non-consanguineous mating. However, with alkaptonuria, which is a very rare condition, one in three parents are first cousins. In such situations the condition may present as a dominant pattern on the pedigree chart.

3. Small families. Some families might only have one or two children. If both parents carry a recessive gene for a particular condition (both are heterozygous) then the offspring have a one in four chance of being affected. Both offspring might be unlucky in that they both inherited the condition, or lucky in that neither of them did. Small family size has to be taken into consideration as it could result in incorrect interpretation of the pedigree of the disease.
4. Non-inherited traits. Not all conditions are inherited. Other factors can cause disease, such as viral infections or exposure to disease-causing agents. In this situation the pedigree chart should not demonstrate any pattern of inheritance that is consistent with genetic principles.

SUMMARY

- Pedigree charts are used to document family history in a pictorial form in order to detect patterns of inheritance.
- Different patterns of inheritance exist depending on whether the gene is autosomal, sex linked, recessive or dominant.
- Limitations include family size, consanguineous mating and the varied effects expressed by the same gene.

FURTHER READING

Bennett, R.L. (1999) *The practical guide to the genetic family history*. New York: Wiley

This is a comprehensive guide on how to take a family history. It is an American text that is mainly aimed at the American market rather than the European market. However, it is very detailed and is a good encyclopaedic reference.

Bennett, R.L., Steinhaus, K.A., Uhrich, S.B., O'Sullivan, C.K., Resta, R.G., Lochner-Doyle, D., *et al.* (1995) 'Recommendations for standardized human pedigree nomenclature'. *Journal of Genetic Counseling*, 4(4): 267–79

Bennett, R.L., Steinhaus French, K., Resta, R.G. and Lochner-Doyle, D. (2008) 'Standardizing human pedigree nomenclature: Update and assessment of the recommendations of the National Society of Genetic Counselors'. *Journal of Genetic Counseling*, 17: 424–33

The two articles above contain all the standardised symbols for pedigree charts. These are the standardised symbols that should be used by all health professionals globally.

The National Genetics Education and Development Centre.
www.geneticseducation.nhs.uk
This is an excellent online resource that provides a series of factsheets on taking and drawing family histories.

08

CLINICAL APPLICATIONS

LEARNING OUTCOMES

The following topics are covered in this chapter:

- screening and testing;
- gene therapy;
- pharmacogenetics.

SCREENING AND TESTING

Genetic screening can be performed at any stage of life. There are a variety of different screening tests available, ranging from the establishment of a family history to quite sophisticated molecular tests.

Prenatal screening and testing

Prenatal screening indicates whether the foetus is at a high or low risk of having a genetic disorder, not if the foetus is affected. Screening is usually done by ultrasound or minimally invasive techniques such as maternal blood tests.

However, a diagnostic test will accurately diagnose if the foetus has a genetic disorder. Small amounts of foetal cells are examined in order to detect either chromosomal abnormalities or DNA mutations. Only a small number of disorders can be diagnosed as genetic testing is not yet available for most single gene disorders.

Samples of foetal cells can be obtained in one of three ways.

1. Amniocentesis

The amniotic fluid surrounding the foetus contains some foetal cells. A small sample of fluid is taken by inserting a needle through the maternal abdominal wall. The foetal cells are examined and the fluid is tested for certain components such as alphafetoprotein, an elevation of which can indicate spina bifida. This test is performed between 15 and 18 weeks' gestation, and it carries a 0.5 per cent risk of miscarriage.

2. Chorionic villus sampling

This involves the removal of cells from the edge of the placenta (the chorion). A small biopsy is obtained by passing a catheter through the cervix of the uterus. Most of the chorion is foetal tissue and is therefore identical to the foetal cells. This test is performed between 10 and 12 weeks' gestation, and it carries a 1 to 2 per cent risk of miscarriage.

3. Foetal blood sampling

A blood sample is taken from the umbilical vein under ultrasound guidance. This test carries a 2 per cent risk of miscarriage.

Neonatal screening

Blood spot samples are taken routinely from the heels of newborn babies between the ages of 5 and 8 days. The dried blood samples are then examined in a laboratory for a small number of genetic conditions. This is known as the Guthrie test. The genetic conditions that all newborn babies are screened for in this way are:

- Phenylketonuria;
- Congenital Hypothyroidism;
- Cystic Fibrosis;
- Medium Chain acyl-CoA Dehydrogenase;
- Sickle Cell disease (only for babies identified as being at risk).

These genetic tests are for conditions with symptoms that, if treated early, can be minimised and controlled.

Carrier testing

This is done to identify individuals who carry a recessive gene. The individual's genome is examined in order to determine if the recessive gene is present. This test is only offered if there is a family history of a genetic disorder. It is of most benefit if both parents are tested as the probability of having an affected child could then be established.

Predictive testing and presymptomatic testing

Predictive testing is done to try to identify individuals who may be at an increased risk of developing a certain genetic condition. This is to identify a susceptibility to that condition, but it is not a firm diagnosis of whether the individual will develop that genetic condition. Presymptomatic testing, on the other hand, is done to detect a genetic disorder, the effects of which might be seen in later life. The individual's genome is examined to determine if they either carry a dominant gene or are homozygous recessive.

The limits of screening and testing

Genetic tests are not available for every disorder and, even if a test is available, results can be misinterpreted. If the presence of a mutated gene has been established, it does not necessarily mean that the individual concerned will develop the disorder in question, as some genes are multifactorial (polygenic genes). If a genetic condition has been identified, the next step would naturally be to offer treatment so that the individual can manage their condition, but is there treatment available?

ACTIVITY 8.1

Huntington's disease is a genetic condition resulting from the presence of a dominant Huntington gene. The presence of just one of these genes means that the individual will develop this condition at some point in their life. The gene has 100 per cent penetrance, which means that the individual will certainly develop the condition.

- Do you feel that it is right to offer a predictive or presymptomatic test in such conditions where there is no cure?
- Do you feel that it is right to deny such a test?

ACTIVITY 8.2

There are many 'self-testing kits' available on the internet. What problems do you think are associated with these tests?

GENE THERAPY

Gene therapy involves inserting copies of a functioning gene into the chromosomes of an individual who carries the faulty gene. There are several different types of gene therapy.

1. **Gene augmentation therapy:** a functional copy of a deleted gene is added into the genome. This is only suitable if the effects of the condition/disease are reversible.
2. **Gene inhibition therapy:** a gene which codes for a protein that inhibits the expression of the mutated gene or interferes with the activity of its product is introduced into the genome.

3. **Killing of specific cells:** this is suitable for cancer cells. Two different types of this therapy exist:
   - The introduction of a 'suicide gene' into the genome. The product of such a gene is toxic to the cell and causes cell death. Care must be taken to avoid widespread cell death.
   - The introduction of a gene that produces a protein that in turn makes the cell vulnerable to attack from the immune system.

Somatic and germ line gene therapy

Somatic gene therapy involves corrective treatment to body tissue. The results would benefit the individual but would not be passed on to future generations.

Germ line gene therapy involves DNA transfer to the cells that produce either eggs or sperm. The results of this therapy would therefore be passed on to future generations. The idea of germ line gene therapy is quite controversial. Although it would prevent future generations from inheriting a particular genetic disorder, there are also arguments against it:

- the effects are unpredictable;
- further defects could be introduced;
- there may be long-term side effects;
- denial of the human rights of future generations;
- potential abuse ranging from designer babies to whole scale eugenics.

ACTIVITY 8.3

Is somatic gene therapy more or less ethical than germ line therapy?

Is gene therapy appropriate for all conditions?

There are many conditions that are suitable for gene therapy. However, there are also some conditions that are not. A suitable condition has to fit the rules of suitability. Rules for suitability are:

1. the condition must be due to mutations in one or more genes;
2. the gene involved must be identified;
3. appropriate knowledge must be available regarding
   - which tissues are involved;
   - what role the encoded protein has;
   - how the mutations in the gene affect the protein's function.
4. a normal copy of the gene will restore normal function of the affected tissue;
5. the corrected gene can be delivered to the cells of the affected tissue.

If all of the rules above apply then the condition would be a good candidate for genetic therapy.

ACTIVITY 8.4

What are the benefits of gene therapy?

The transfer of genes into cells for gene therapy

Usually a gene cannot be directly inserted into a cell. The gene must be delivered using a vector. A vector is a carrier that transports the gene into the cell and into the genome. The most commonly-used vectors are viruses, although liposomes (fatty particles) can also be used.

Different vectors behave differently in their ability to transfer genes. The suitability of the vector depends upon the specific characteristics and requirements of the gene and the identified condition (see Table 8.1).

Preparing a vector in the laboratory

Vectors are prepared in three steps.

Step 1: building the vector

Virus DNA exists in a circular form called a plasmid. The replication gene is removed from the virus DNA as well as the cap gene (this codes for the shell that surrounds the virus particle) (see Figure 8.1).

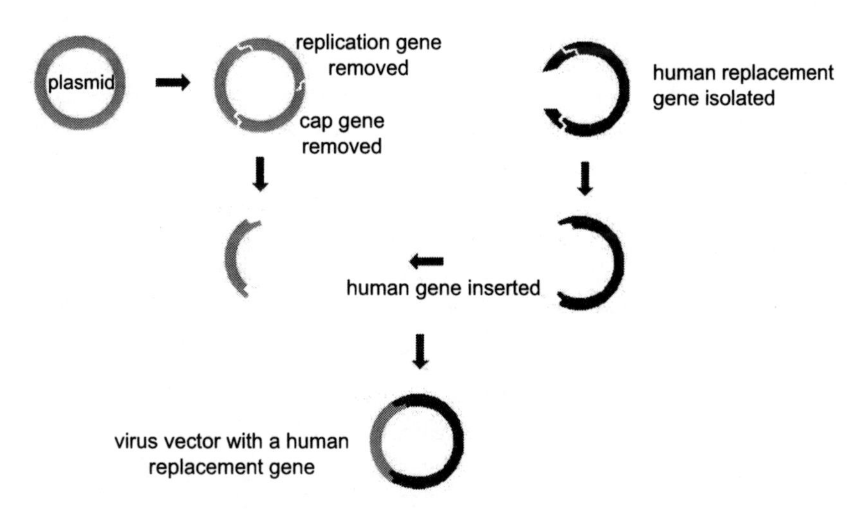

Figure 8.1 *Building the vector*

Table 8.1 *Vector types*

Name	Adenovirus	Retrovirus	Adeno-associated virus
Effect on humans	Common cold virus.	e.g. HIV.	Does not cause illness in humans.
Type	Viral vector.	Viral vector.	Viral vector.
Target	Both dividing and non-dividing cells. It is possible to target specific cell types by altering the proteins on the virus surface to recognise special proteins on the target cell's surface. Maximum length of DNA that can be inserted is 7,500 base pairs.	Infects only dividing cells. Maximum length of DNA is 8,000 base pairs.	Wide range of dividing and non-dividing cells. They are unique in that they need to have a 'helper' adenovirus with them to infect cells. Maximum length of DNA is 5,000 base pairs.
Activation	Once it infects cell, travels to cell's nucleus where its genes are activated.	Once it infects a cell, travels to the nucleus. Carries its genetic material in the form of RNA, which must then be converted to DNA before some genes can be activated. The retrovirus contains enzymes that do this job.	Once it infects a cell, it travels to the nucleus where its genes are activated.
Integration	Will not integrate into the host cell's genome. After 1 or 2 weeks, the cell will discard it and gene activation will be lost.	Integrates into host cell's genome in random locations. Once integrated, it will be duplicated along with the rest of the DNA when the cell divides.	Will integrate into the host cell genome 95% of the time. It will integrate into a specific region on chromosome 19, greatly reducing the chance that integration will disrupt the function of other genes in the cell.
Side effects	Can cause an immune response. This can be reduced by removing proteins that trigger this response.	As it integrates randomly into the genome, there is a chance that it will integrate into a place where it will disrupt another gene. Can also cause an immune response.	Does not normally cause an immune response.

Name	Herpes Simplex virus	Liposome	Naked DNA
Effect on humans	Oral and genital herpes.	A plasmid DNA that is packaged into liposomes (miniature lipid-based packets which are similar to the cell's own membrane).	A lone plasmid DNA molecule. Some cells will bind and take up naked DNA.
Type	Viral vector.	Non-viral vector.	Non-viral vector.
Target	Infects cells of the nervous system. Maximum length of DNA is 20,000 base pairs.	Not specific to any cell. Enters cells less efficiently than viral vectors.	Not specific for any cell type. Enters cells less efficiently than viral vectors.
Activation	Carries genetic material as a single stranded DNA. Once in the host cell it travels to the nucleus where its genes are activated.	Upon entering the cell, plasmid DNA is transported into the cell's nucleus, where its genes will be activated.	Upon entering the cell DNA is transported to the nucleus where genes are activated.
Integration	Although will not integrate into host's genome, it can stay in the nucleus for a very long time as a circular piece of DNA that replicates when the cell divides. It will not disrupt the function of other genes in the host cell.	Effectiveness at integration is poor.	Will not normally integrate unless engineered to do so. Even when engineered it is not always effective.
Side effects	Can cause an immune response.	Will not generate an immune response, but some types are known to be toxic.	Will not generate an immune response. Generally not toxic.

This results in a virus vector that contains the identified human replacement gene and not the original virus genes involved in virus replication.

Step 2: producing the virus

The vector is then transferred into a cell line medium. Adenoassociated viruses can not replicate on their own so another type of virus, the adenovirus, is added to the cell line medium. The removed replication gene and the cap gene from the adenoassociated virus are also added to the cell line medium (see Figure 8.2).

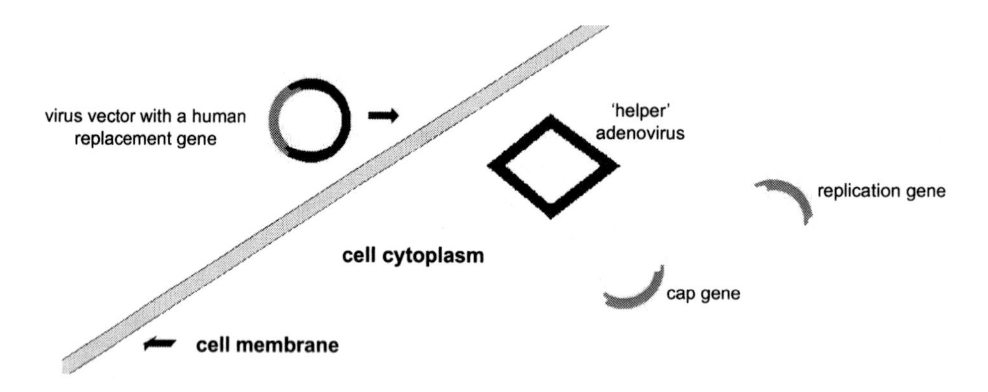

Figure 8.2 *Producing the virus*

The cells within the cell line medium combine the new plasmid with cellular proteins to produce new viruses. Cells that are full of viruses break open and release the viruses into the growth medium.

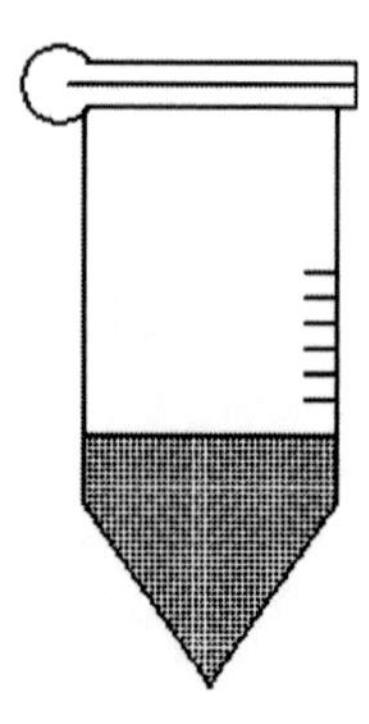

Figure 8.3 *Altered viruses in growth medium*

Step 3: infecting the body cells

The vector can either be given directly into a specific tissue in the body, where it is taken up by individual cells or a sample of cells can be removed from the individual and exposed to the vector in a laboratory setting. These cells are then returned to the body with the vector's DNA within the cell nucleus.

Ex vivo:	cells are removed and repaired in a laboratory setting. Cells are then reimplanted into the body.
In situ:	the repairing gene is introduced directly into the defective tissue/organ.
In vivo:	the repairing gene is administered to the individual where it eventually reaches the exact point needed.

Gene therapy challenges

There are many challenges involved with gene therapy.

1. **Transport method**: many vectors are damaged or destroyed by the body's immune system. There is a need to bypass several cell recognition systems before the DNA can be inserted into the genome.
2. **Right cell**: there could be a risk of the vector affecting non-target cells within the body. This poses a major problem with regards to 'suicide genes' or it could even alter germ line cells.
3. **Right place**: the new DNA might be inserted into the wrong position on the genome. This could result in a mutation, even leading to cancer.
4. **Longevity**: once the introduced DNA is correctly integrated into the correct position in the genome, it needs to work in harmony with all the other genes. In many clinical trials, the effects are very short-lived.

Gene therapy has existed in experimental form since the late 1980s. Although great progress has been made there have also been major set-backs. Experiments conducted in laboratories have not always been successfully replicated in human subjects. The main problem for gene therapy has been basic biology. It is a great challenge to insert new genes into billions of target cells within the body. If this hurdle is overcome, then the new gene needs to be able to function. Quite often these new genes are suppressed or even destroyed. These problems are yet to be overcome and this is the reason why gene therapy is only available in clinical trials, and not as a general treatment option (see Case study 8.1).

CASE STUDY 8.1

Cystic Fibrosis – an example of gene therapy

Problem: symptoms include abnormal mucus secretions that can lead to an increased risk of infections.

Genes: mutation in CFTR (Cystic Fibrosis Transmembrane Regulator) gene on chromosome 7. The mutated gene consists of 4,440 base pairs, while the normal gene has 4,443 base pairs (one codon missing in the mutated gene). The codon TTT which codes for phenylalanine is missing from the mutated gene.

The normal CFTR gene codes for functioning chloride ion channels in cell membranes. Normally, cells have more chloride ions outside the cell than they do inside the cell as the chloride ion channels enable the cell to pump chloride out of the cell. This leads to the attraction of more water outside the cell into the mucus layer. The mucus layer traps debris and bacteria and the cilia sweep away the harmful debris. The mutated gene codes for faulty chloride ion channels that cannot transport chloride ions across the cell membrane. The mucus layer outside the cell becomes thick and sticky and is not easily removed. This then increases the risk of infections.

Cystic Fibrosis – is it a good candidate for gene therapy?

- **Does the condition result from mutations in 1 or more genes?**
  Yes – an inherited, recessive disorder.
- **Which gene is involved?**
  Single gene: CFTR on chromosome 7.
- **Disorder**
  Characterised by accumulation of sticky mucus in lungs and digestive system.
- **Will adding a normal gene correct the condition?**
  Yes.
- **Is it feasible to deliver the gene to cells of the affected tissue?**
  Yes, in part. Lungs would be feasible probably by the use of inhalers; the digestive system is less accessible.

Choosing a vector for Cystic Fibrosis (see Table 8.1)

- Every viral vector has an upper size limit for the inserted gene. The CFTR gene is 4,443 base pairs long. As the gene is small, none of the vectors can be excluded based on gene length.

- No virus can infect every type of cell. For Cystic Fibrosis the target is airway cells. The Herpes Simplex virus cannot be used as it only infects cells of the nervous system.
- Some viruses can only infect cells that divide regularly. Airway cells divide infrequently. The retrovirus can be excluded at this stage as it can only infect dividing cells.
- To effect any change in the patient's lung tissue, the vector needs to enter a high proportion of the cells. Liposomes and naked DNA can be excluded here as they enter less efficiently when compared with viruses.
- Some virus vectors can cause an immune response in the patient. Adenovirus can be excluded due to the risk of causing an immune response.
- By using the exclusion criteria above, the best choice for Cystic Fibrosis gene therapy is the Adeno-associated virus.

Clinical trials on patients with Cystic Fibrosis

1993: the first clinical trials began. Adenovirus vector was used.

Why did it not work?
It appeared at first that the virus had entered the cells and that the CFTR gene was working. Later trials with different patients showed that the level of the CFTR gene was too low.

Initial thoughts
Adenovirus can not easily enter cells. Higher doses were then tried, which resulted in an immune response.

Conclusion
Scientists needed to find a way to improve the ability to enter cells and to reduce the chances of an immune response.

1995: liposomes were chosen as a vector.

Why did it not work?
Low levels of the CFTR gene were delivered to the target cells. The DNA that was delivered was not integrated into the genomic DNA. Activity dwindled over a short period of time. Side effects included fever and inflammation.

Conclusion
Mildly effective but the effect was very short-lived.

1998: Adeno-associated virus was used as a vector.

Why did it not work?
Although there was no immune response or adverse side effects, the vector did not enter cells effectively.

Conclusion
More scientific information needed on how this virus affects cells. Gene therapy for Cystic Fibrosis is still at the experimental stage.

ACTIVITY 8.5

What factors have kept gene therapy from becoming an effective treatment?

ACTIVITY 8.6

What are the ethical considerations for using gene therapy?

PHARMACOGENETICS

The human genome project has established that 99.9 per cent of human genes are identical between one person and another. The remaining 0.1 per cent of the genes in the human genome accounts for the variation between individuals. These variations or polymorphisms do not normally affect growth and development but may contribute towards an individual's susceptibility to disease or affect the ways in which drugs are metabolised in the body.

Pharmacogenetics, or pharmacogenomics, is a field of study that has arisen from the study of how genes affect drug interactions within the body. Although the terms 'pharmacogenetics' and 'pharmacogenomics' are sometimes used interchangeably, they are terms that indicate the approach taken to the study of genes and drug interactions.

Pharmacogenetics

An unexpected drug response leads to the search for a genetic cause. Studies in this area involve identifying genes that affect an individual's response to current drugs.

Pharmacogenomics

Genetic differences are identified first that might explain different responses to a drug. Studies involve identifying specific genes and gene products, associated with various disorders, which may act as new drug targets.

For the purpose of this chapter the term pharmacogenetics will be used.

The aim of pharmacogenetics is to understand how genetic variation affects drug interactions within the body. This knowledge can then lead to the tailoring of drugs that are best suited to that individual. This is known as **personalised medicine**.

Variations in genes can result in variations in the clinical response to drugs. Genes code for drug receptors, drug transporters, cell signalling pathways and enzymes, all of which are involved in drug metabolism and disposition. Variations in these genes can lead to an exacerbated drug effect or to the lack of a response to a particular drug. A standard drug regime can therefore induce inappropriate drug concentrations at the site of action within the body.

ACTIVITY 8.7

Make a list of all the benefits of personalised medicine. Think of the benefits to the patient and the service provider.

New drug targets

Many of the proteins coded for by the polymorphic genes can become targets for new drugs. Several polymorphic genes have been identified in the human genome. However, the function of each polymorphism, its product and its variant forms are yet to be identified in many genes. A further challenge is then to establish whether the gene product is of any pharmacological importance.

There are some important developments in the recognition of the variation in the number of receptor sites produced by the genes. Receptor sites are composed of proteins on the cell membrane that have a binding effect with certain chemical messengers. Once the chemical messenger has combined with the receptor site, the receptor protein initiates a chain of events within the cell. The variations have shown that some people produce more receptor sites in and on cell surfaces compared to others. See the following example regarding HER2 receptors in breast cancer (Case study 8.2).

CASE STUDY 8.2

HER2 receptors and breast cancer

In primary breast cancer, 25 to 30 per cent of patients over-express a cancer-related gene called HER2. The HER2 gene produces a receptor protein on the surface of breast cells. This receptor is thought to have an important role in cell division. When the HER2 gene is over-expressed, extra receptors are formed that lead to the cells dividing at an uncontrolled rate, and then becoming cancerous. The drug Trastuzumab (Herceptin) is an artificially developed antibody that blocks some of the receptor sites. This has the effect of limiting the amount of cell division and thereby preventing the growth of the cancer.

Drug response

Polymorphic genes can lead to either a lack of a therapeutic effect from a drug or an exacerbated response. These subtle genetic variations between individuals mean that the dose at which a drug will be effective will vary enormously between one individual and another. See Case study 8.3 regarding the P450 enzymes.

CASE STUDY 8.3

The P450 enzymes

An important set of drug-metabolising enzymes is the P450 enzymes found in the liver. Three of these enzymes, CYP 2D6, CYP 2C9 and CYP 3A4, are especially important in metabolising drugs. The CYP 3A4, which metabolises about 50 per cent of all prescribed drugs, does not appear to be polymorphic. However, the CYP2D6, which accounts for the metabolism of about 25 per cent of all prescribed drugs, has been found to be polymorphic.

Variations in the information contained in the CYP 2D6 gene determine how much of this enzyme is produced in the liver. Individuals who produce low levels of CYP 2D6 will require smaller doses of the drugs that are metabolised by this enzyme. Individuals who have a lot of this enzyme metabolise drugs quickly and will need larger doses to get the same therapeutic effect. The drug codeine is metabolised by the CYP 2D6 enzyme and up to 10 per cent of the population do not get any pain relief from codeine due to this polymorphism. The lack of effectiveness, in some individuals, of the antidepressant Prozac is also thought to be due to this polymorphism. There are many identified drugs that are metabolised by the CYP 2D6 enzyme, a large proportion of which are drugs used for treating cardiovascular, neurological and psychiatric conditions where the therapeutic range is quite narrow.

The CYP 2C9 enzymes are highly polymorphic but only account for the metabolism of about 5 per cent of all prescribed drugs. The anticoagulant warfarin is metabolised by the CYP 2C9 enzyme. Individuals with polymorphisms that reduce the enzymes' activity need to have reduced doses of warfarin to avoid life-threatening bleeding.

The benefits of pharmacogenetics

Some potential benefits include:

- **right drug** – drugs could be developed to target specific conditions, limiting the damage to healthy tissues.
- **right dose** – instead of just using age and weight as dosage indicators, an individual's genetics would be taken into account. This would reduce the chances of an overdose.

- **right patient** – prescribers would have a clearer idea of which drug to use based on the patient's genetic profile. This would eliminate the need for a trial and error approach to drug therapy.

ACTIVITY 8.8

How do you think pharmacogenetics can affect the quality of health care?

The limitations of pharmacogenetics

Some identified limitations include:

- **Lack of knowledge** – although polymorphisms have been identified since the completion of the human genome project, the knowledge of what proteins these genes code for is as yet incomplete. Once the protein of a polymorphic gene has been identified, further research is then needed to establish its pharmaceutical relevance.
- **Lack of resources** – several genomics companies have produced tests to identify a limited amount of polymorphisms. These tests, however, are only currently employed in large teaching hospitals or in specialist centres. They are not widely available as yet.
- **Lack of regulation** – currently, in the UK, the efficacy of medicines is assessed by the Medicines and Healthcare Products Regulatory Agency (MHRA). Regulation and quality of genetic tests is also the responsibility of the MHRA. Regulations relating specifically to pharmacogenetics have not yet been established. However, with the development of pharmacogenetics in practice, guidelines need to be established regarding prescribing, testing and usage labels.

ACTIVITY 8.9

What are some of the challenges that face pharmacogenetics?

SUMMARY

- **Screening and testing**
  - Screening can be used to establish the probability of a genetic disorder in an individual.
  - Testing is employed to accurately establish the presence of an abnormal gene.
  - Testing techniques are not available for every genetic condition.
- **Gene therapy**
  - Involves inserting a functioning gene into the genome.
  - Gene therapy is not suitable for all genetic conditions as yet.

- Different types of vectors, chosen for their suitability and efficiency, are used to transport the gene into the genome.
- Gene therapy is not available as yet as a mainstream treatment option as it is still in the research stage.

- **Pharmacogenetics**
  - Examines how genes play a role with drug interactions within the body.
  - Personalised medicine involves tailoring of drugs and their dosage levels according to the individual's genetic makeup.

FURTHER READING

Human Genetics Commission. **www.hgc.gov.uk**
The Human Genetics Commission, as the UK advisory body on new developments in genetics, has excellent web links to over 60 UK and non-UK based organisations.

National Genetics Education and Development Centre
www.geneticseducation.nhs.uk/teaching-genetics/pharmacogenetics.aspx
The pharmacogenetics site of the National Genetics Education and Development Centre is well worth looking at as it has a wealth of information and the site is still growing. It also has good links to pharmacogenomics web resources.

Nuffield Council on Bioethics. **www.nuffieldbioethics.org**
The Nuffield Council have released many reports on the ethical considerations of the advancements in genetics. The *Report on genetic screening* (1993 and supplement 2006) as well as *Pharmacogenetics: Ethical issues* (2003) sets out recommendations and advice for UK policy makers.

Parliamentary Office of Science and Technology (2004) NHS Genetic Testing.
www.parliament.uk/documents/post/postpn227.pdf
This is a short document outlining the current use of genetic tests and possible future applications (including pharmacogenetic testing) in the UK.

UK Genetic Testing Network. **www.ukgtn.nhs.uk/gtn/Home**
This site provides information about which genetic tests are available in the UK. You can search this database by region or by disease.

UK Screening. **www.screening.nhs.uk**
This site explains what screening is and identifies its limitations. It also provides access to UK screening policies and identifies the differences between regional areas within the UK. It is the UK screening committee that recommends which screening programmes should be used in the UK.

09

CANCER GENETICS

LEARNING OUTCOMES

The following topics are covered in this chapter:

- cancer-causing genes – oncogenes and tumour-suppressor genes;
- the differences between oncogenes and tumour-suppressor genes;
- environmental triggers that can cause carcinogenic gene mutations;
- familial cancers;
- genetic causes of common cancers.

INTRODUCTION

The transformation of normal cells into cancerous cells occurs following mutations within the cancer-causing genes. Cell division is normally under strict control, which includes when a cell should divide and how often. Any mutation that affects cell division can give rise to a tumour. Mutations in the DNA coding responsible for cell division can arise as a result of an environmental agent (such as tobacco smoke) or from a familial disposition. Quite often both are involved. There are broadly two types of cancer-producing genes: oncogenes and tumour-suppressor genes.

ONCOGENES

Oncogenes are mutated forms of normal genes called proto-oncogenes. Proto-oncogenes control cell division and cell differentiation. Proto-oncogenes are only active when the cell needs to divide. When a proto-oncogene mutates into an oncogene it becomes permanently 'switched on'. This leads to uncontrolled cell division and cell growth. Normally only one of a pair of proto-oncogenes needs to be in the mutated oncogene form for a tumour to develop, as oncogenes act as dominant genes.

There are over 100 recognised oncogenes in the human genome, many of which affect different stages of cell division and cell growth. These oncogenes can be classified into the following categories.

1. Growth factors

Growth factor oncogenes produce factors that stimulate cell growth. An example is the **SIS oncogene** that leads to the over-production of platelet-derived growth factor which stimulates cell growth (Case study 9.1).

CASE STUDY 9.1

Meningioma

Meningioma is a primary tumour arising from the cells that surround the arachnoid villi in the meningial layers of the central nervous system. It is a slow-growing tumour that normally presents in middle age.

The gene associated with meningiomas is the SIS gene, which is also known as the PDGFB gene (Platelet-Derived Growth Factor Beta Polypeptide). This gene is situated on the long arm of chromosome 22 (22q) and normally acts as a growth regulator. A mutation in this gene can lead to uncontrolled cellular growth.

2. Growth factor reception

These receptors on the cell surface can become over-sensitive to growth factors by being permanently 'switched on'. An example is the **HER-2/neu oncogene.** This gene is found in about 30 per cent of all breast cancers (Case studies 9.2 and 9.3).

CASE STUDY 9.2

Breast cancer

About 30 per cent of all cases of female breast cancer display an abnormally large amount of HER-2/neu surface receptors on the cancerous cells. These receptors function as growth factor receptors that stimulate cell division. When the proto-oncogene HER-2/neu becomes mutated to the HER-2/neu oncogene, this leads to the increased number of cell growth factor receptor sites. This in turn can lead to increased cell division and eventually to malignancy.

CASE STUDY 9.3

Multiple Endocrine Neoplasia Type 2

This condition often results in malignant growths in the thyroid gland and in the adrenal glands, which results in very high blood pressure. Multiple endocrine neoplasia results from a gain in function mutation in the RET gene on chromosome 10 (10q). The RET gene encodes for a protein that is involved in signalling within cells. Mutations in this gene over-activate the protein's function, which triggers cell growth and division leading to the development of a tumour.

3. Signal transducers

This involves the pathway between the growth factor receptors and the nucleus of the cell. An oncogene involving signal transducers means that this pathway is always 'switched on'. An example is **ABL gene** abnormalities that can lead to the development of Chronic Myeloid Leukaemia (Case study 9.4). Also, **RAS gene** abnormalities have been found in most cancers including lung, ovarian, colon and pancreatic cancer.

CASE STUDY 9.4

Leukaemia

Chronic Myeloid Leukaemia (CML) affects blood cells. Malignant leukaemic cells replace the normal blood cells in the bone marrow. CML is usually caused by the translocation of chromosomal material. Translocation occurs between parts of the long arm of chromosome 9 (9q) and the long arm of chromosome 22 (22q).This translocation brings together two genes.The ABL proto-oncogene on chromosome 9 is inserted onto the BCR gene (Breakpoint Cluster Region gene) on chromosome 22. This results in the formation of a new BCR-ABL gene. The BCR-ABL gene codes for a protein that activates the signal transduction pathways, which leads to uncontrolled cell growth.

The reciprocal translocation between 9q and 22q leaves a characteristic modified version of chromosome 22 that has been called the 'Philadelphia chromosome' after the city where it was first discovered.

4. Transcription factors

This is the final stage of events that lead to cell division. Transcription factors act directly on the nuclear DNA and control which genes are active in producing proteins. An example is the **MYC gene** which becomes overly active and stimulates cell division. Lung cancer, leukaemia and Burkitt Lymphoma have been attributed to an over-active MYC gene (Case study 9.5).

CASE STUDY 9.5

Burkitt lymphoma

Burkitt lymphoma is a rare cancer that affects mostly children. It is a B lymphocyte cancer of the jaw. This type of cancer arises from chromosome translocations that affect the MYC gene on chromosome 8. Mutations in the MYC gene act as transcription factor oncogenes. Translocations between chromosome 8 and chromosomes 2, 14 and 22 have all been recorded.This changes the effect of the MYC gene as a proto-oncogene that is involved in the regulation of cell growth and division.The altered MYC gene results in an oncogene that becomes involved with the over-production of cells, which in turn become cancerous.

5. Cell death regulators

These oncogenes prevent the cell from committing suicide. This leads to abnormal cell growth that can become cancerous. An example is the **BCL2 oncogene** that is often activated in lymphoma cells (Case study 9.6).

CASE STUDY 9.6

B Cell Lymphoma (Follicular Lymphoma)

There are many different types of B cell lymphomas, with follicular lymphoma being one of the most common types. B cells are lymphocytes that are involved in the humoral immune response by producing antibodies. Malignancies in these cells arise from the lymphatic system.

The BCL2 gene has been associated with follicular lymphoma. Translocation between chromosome 14 and chromosome 18 results in the over-expression of the BCL2 gene. The gene is normally situated on chromosome 18 but, after translocation to chromosome 14, the gene becomes over-expressed.This gene is normally involved in preventing apoptosis (cell death) and over-expression leads to the cells becoming 'immortal'. The extended life of the cells then leads to the increased risk of developing DNA replication errors.

ACTIVITIES 9.1 AND 9.2

1. When are proto-oncogenes normally active?
2. Genes occur in pairs. Do both proto-oncogenes need to be in a mutated oncogene form for a tumour to develop?

TUMOUR-SUPPRESSOR GENES

These are genes that slow down cell division, repair DNA errors and instruct the cell when to die (apoptosis). Mutations in tumour-suppressor genes can lead to uncontrolled cell division with DNA errors that can lead to cancer. Mutations in these genes can result in loss of function. As genes occur in pairs, the loss of function in one gene does not result in cancer as the individual will have another functioning gene. Tumour-suppressor genes will usually only lead to cancer if both genes have loss of function through mutation as these genes normally act as recessive genes. There are over 30 recognised tumour-suppressor genes in the human genome. These genes can be classified into three types.

1. Cell division control genes

Some tumour-suppressor genes control cell growth and cell division. Cell growth and cell division may only accelerate if both genes are in a mutated form. An example is the **RB gene**, which is an inhibitor of the cell cycle. Mutations in this gene have been found in retinoblastoma, bone cancer, bladder cancer and breast cancer (see Case studies 9.7 to 9.11).

CASE STUDY 9.7

Multiple Endocrine Neoplasia Type 1

This condition results in the development of malignant tumours in the endocrine organs. The most commonly involved organs are the thyroid gland, parathyroid, pancreas and the anterior pituitary. Gastrointestinal tumours have also been linked with this condition. The age of onset is variable, with about half of cases developing by the age of 20 years. All individuals with this condition will have developed tumours by the age of 50 years.

Multiple endocrine neoplasias arise as a result of defects in the MEN1 gene, which encodes for the protein called menin. This gene is situated on the long arm of chromosome 11 (11q). Mutations in both copies of the MEN1 gene lead to uncontrolled cell growth and division. The cells divide at an accelerated rate, giving rise to a tumour.

CASE STUDY 9.8

Neurofibromatosis

This condition results in tumour-like growths in skin and in the nervous system. There are two different classes of neurofibromatosis, type 1 and type 2, both of which are autosomal dominant conditions. Type 2 is an example of a tumour suppressor cell division control gene.

- **Type 1 (von Recklinghausen disease)** – type 1 is the most common type of neurofibromatosis.The disease is characterised by a number of different conditions, not all of which have to be present for a positive diagnosis. Conditions include pigmented skin lesions, neurofibromas, tumour of the optic nerve and benign hamartoma of the iris. The risk of an individual with type 1 neurofibromatosis developing a malignancy is around 5 per cent. The gene thought to be responsible for this condition is the NF 1 gene that is located on the long arm of chromosome 17 (17q), the exact function of which is not entirely clear.
- **Type 2** – this disease is characterised by the development of benign tumours on both auditory nerves. However, it can also result in the development of malignant tumours in the central nervous system involving the Schwann cells, glial cells or the meningial cells. The neurofibromatosis type 2 gene has been named Merlin and can be found on the long arm of chromosome 22 (22q). This gene acts as a tumour suppressor gene. Mutations in both copies of the Merlin gene can result in neurofibromatosis type 2.

CASE STUDY 9.9

Retinoblastoma

Retinoblastoma is a malignant tumour that affects the cones of the retina. It is a tumour that affects children. Both hereditary and non-hereditary forms exist. Hereditary forms of retinoblastoma often result in multiple tumours in both eyes, whereas non-hereditary forms tend to result in one tumour in one eye only.

The retinoblastoma gene, RB 1, can be found on the long arm of chromosome 13 (13q). The RB 1 gene acts as a tumour-suppressing gene by binding transcription factors so that cell division cannot occur. Retinoblastoma occurs as a result of a faulty or missing RB 1 gene.

In hereditary cases, the infant will have inherited one faulty RB 1 gene and one functioning RB 1 gene. When the functioning gene suffers a mutation, deletion or translocation this cell is likely to become malignant. Hereditary retinoblastoma requires two point mutations: one inherited and one sporadic post conception. Non-inherited retinoblastoma requires two somatic mutations post conception in the RB 1 gene.

CASE STUDY 9.10

Von Hippel–Lindau syndrome

This is a multisystem disorder that is characterised by the abnormal growth of capillary blood vessels, which leads to haemangioblastomas. Growths of haemangioblastomas with this syndrome usually occur in the central nervous system, the retina, adrenal glands and the kidneys. Some individuals may also become deaf due to the involvement of the labyrinth in the inner ear.

The gene for this syndrome is the mutated VHL gene, which is situated on the short arm of chromosome 3 (3p). It is inherited in a dominant fashion. The VHL gene is a tumour-suppressor gene that normally stops uncontrolled cellular growth and proliferation. Over three quarters of Von Hippel–Lindau cases have inherited the VHL mutated gene, the remaining arising from *de novo* mutations.

CASE STUDY 9.11

Wilms' tumour

Wilms' tumour is a malignancy of the kidney that occurs in very young children. It is one of the most common solid tumours occurring in childhood.

A mutation in the WT1 gene has been found in children with this cancer. The WT1 gene is situated on the short arm of chromosome 11 (11p). This gene, which normally halts mitosis in the developing kidney tubules, is missing in the foetus. Deletion of DNA bases within the WT1 gene means that the child's kidneys will have some cells that continue to divide at an accelerated rate. Other Wilms' tumour genes have also been identified, including WT2 also on chromosome 11 (11p), WT3 on chromosome 16 (16q), WT4 on chromosome 17 (17q) and WT5 on chromosome 7 (7p). Germ line mutations account for less than 10 per cent of cases.

2. DNA repair genes

All nuclear DNA must be replicated prior to cell division. Sometimes errors occur in the copying process. DNA repair genes 'proof read' the DNA and correct any errors that might have occurred. If the DNA repair genes are faulty then copying errors remain unchecked and mutations occur, including oncogenes and more tumour-suppressor genes. An example is **HNPCC** (Hereditary Non-Polyposis Colon Cancer). This mutated tumour suppressor gene causes early onset colon cancer and endometrial cancer. This gene acts as an autosomal dominant gene (see Case studies 9.12 and 9.13).

CASE STUDY 9.12

Breast cancer

Up to 10 per cent of hereditary breast cancer is due to mutations on the BRCA 1 gene on chromosome 17 and the BRCA 2 gene on chromosome 13. An individual who has a mutation in either of these genes increases their risk of developing breast cancer during their lifetime. Up to 15 per cent of breast cancer cases in men appear to be due to the BRCA 2 gene.

BRCA 1 is a gene thought to be involved in the regulation of the cell cycle and indirectly involved with DNA repair and DNA recombination. BRCA 2 on chromosome 13 is directly involved with other proteins that act as transcriptional regulators.

As tumour-suppressor genes, both copies of BRCA 1 and/or BRCA 2 need to be in a mutated form before a tumour can develop.

CASE STUDY 9.13

Colon cancer

Between 5 and 10 per cent of colon cancers follow the pattern of autosomal dominant inheritance. The remaining 90 to 95 per cent appear sporadic. The two major types of colon cancer are:

a) **Hereditary nonpolyposis colorectal cancer (HNPCC)** – among some of the genes involved in inherited colon cancer are MSH 2 and MSH 6, which are both on chromosome 2, and MLH 1 on chromosome 3. These genes normally aid in the repair of DNA replication errors. Mutated forms of these genes result in DNA errors not being corrected, which can result in cancer of the colon.

b) **Familial adenomatous polyposis (FAP)** – the inheritance of a single mutated tumour-suppressor gene can result in multiple benign polyps in the lining of the colon. The remaining 'normal' tumour-suppressor gene has to undergo mutation for the polyps to progress to a cancerous state. The most common form of FAP is due to mutations of the APC gene on chromosome 5. APC codes as a tumour-suppressor gene as well as controlling the expression of the MYC oncogene.

3. Cell suicide genes

Prior to cell division, if the DNA contains too many copying errors or is damaged in some way, a protein called **P53** increases in activity and stimulates the production of a protein called **P21**. The P21 protein prevents the cell from dividing. P53 can also initiate apoptosis (cell death) if the damage to the DNA is severe.

The **P53 gene** is known as the guardian of the genome as it prevents cells with DNA replication errors from dividing. A defective P53 gene will result in the unrestricted replication of cells that contain numerous mutations, which can then lead to cancer. The P53 gene appears to be subject to mutation as more than half of human cancers do not have a functioning P53 protein (see Case study 9.14).

CASE STUDY 9.14

Lung cancer

The two main types of lung cancer are small cell lung cancer and non-small cell lung cancer. A number of mutated genes have been identified in numerous lung cancer cases.

- **Non-small cell lung carcinoma** – this is by far the most common type of lung cancer. It is also caused by the mutagenic effects of cigarette smoke. A number of different genes have been identified in causing non-small cell lung carcinoma, including oncogenes as well as tumour-suppressor genes. Germ line mutations in the P53 gene have been closely linked, especially among smokers. No one mutation appears likely to result in a cancerous growth. Mutations in a variety of genes, together with the influence of external factors, appear to be necessary for the formation of a tumour.

ACTIVITY 9.3

When are tumour-suppressor genes active?

The differences between oncogenes and tumour-suppressor genes

One of the most important differences between oncogenes and tumour-suppressor genes is that oncogenes result from being permanently 'switched on', while tumour-suppressor genes cause cancer from being 'switched off'. Oncogenes generally behave as dominant genes as they become overactive but tumour-suppressor genes mainly act as recessive genes due to their loss of function. Another major difference is that the majority of oncogenes arise from the mutations of normal genes (called proto-oncogenes) that occur during an individual's lifetime (acquired mutation). Tumour-suppressor gene abnormalities, on the other hand, can be inherited as well as acquired.

ENVIRONMENTAL TRIGGERS

There are many recognised cancer-causing agents (carcinogens) in today's society. These carcinogenic agents can be broadly categorised into the following three groups.

1. Chemicals

Although most people perceive that most carcinogenic chemicals originate from man-made materials and heavy industry, these only account for a small proportion of chemical mutagens. Natural substances can also be mutagenic with a potential to develop into cancer. Environmental chemical mutagens include some foods, drugs, smoke, paints, pesticides and petro-chemicals. Exposure to such chemicals can be by ingesting, inhaling or by skin contact (see Case study 9.15).

Some chemicals alter the DNA within the cell by substituting DNA base pairs, creating deletions or additions, or even by causing DNA strand breakages and cross-linking.

CASE STUDY 9.15

Lung cancer

The two main types of lung cancer are small cell lung cancer and non-small cell lung cancer (see above). A number of mutated genes have been identified in numerous lung cancer cases.

- **Small cell lung carcinoma** – metastases are usually already present at the time of diagnosis of small cell lung carcinoma. This type of cancer responds to both chemotherapy and radiotherapy, but surgery is usually not indicated as metastases are usually present.

  The deletion of the short arm of chromosome 3 (3p) in one chromosome only has been associated with small cell lung carcinoma. Studies have shown that this chromosomal abnormality is caused by cigarette smoke.This is an example of a chromosomal abnormality that is caused by an external factor.

ACTIVITY 9.4

Can you think of any other known carcinogenic substances?

2. Electromagnetic radiation

Radioactive decay involves the discharge of alpha, beta and gamma rays. The mutagenic effect of these rays depends upon their speed, mass and electric charge. Radiation can cause major deletions, translocations and aneuploidy (see Chapter 6 on mutations).

- **Ultraviolet light** – the effect of ultraviolet light on DNA is that it can lead to the linking of thymine and cytosine bases (T-T or T-C or C-C). This occurs in somatic cells and does not create germ line mutations. It is a major cause of skin cancer (Case study 9.16).

CASE STUDY 9.16

Malignant melanoma

Malignant melanomas arise from melanocytes within the epidermis which become cancerous. The increased incidence of melanomas in recent years suggests a hereditary factor. Germ line mutations in the CDKNA 2 gene on chromosome 9 increase susceptibility to melanoma. Ultraviolet radiation has also been identified as a risk factor. Increased sun exposure has been established as a risk factor for all types of skin cancer. Melanomas are not usually associated with the cumulative ultraviolet exposure but with intense intermittent exposure. It is thought that intense exposure results in substantial damage to the cellular DNA but not in apoptosis. Mutations in the nuclear DNA occur and the cell survives to divide. The risk of melanoma is associated with sun exposure that results in sun burn.

The CDKNA 2 gene normally codes for a protein called P16. The P16 protein acts as an important regulator in cell division as it stops DNA synthesis before cell division. Mutations in the CDKNA 2 gene result in the cells continuously dividing.

- **Atomic radiation** – there are four different types of atomic radiation:
  - **Natural**: the levels of natural background radiation vary from area to area. The most abundant natural background radiation is radon, which accounts for over half of all natural background radiation.
  - **Cosmic**: the intensity of cosmic rays increases with altitude. Some studies have shown that airline pilots are at an increased risk of developing leukaemia because of the mutagenic effects of cosmic rays.
  - **Man-made**: radiation workers in nuclear power stations and those exposed to fallout in nuclear testing sites are most at risk. Bone-surface-seeking isotopes are a major risk to developing leukaemia.
  - **X-rays**: radiation waves from X-rays can alter DNA in two ways. DNA can become directly mutated by ionising impact or free radicals can be created that act on DNA. Free radicals can travel in the blood stream and affect cells that were not originally exposed to the X-ray.

ACTIVITY 9.5

There has been much discussion in the media regarding sun-beds and how dangerous they are to health. How can sun-beds cause cancer?

3. Viruses

Viruses have been causally linked to cancer. Viruses can 'modify' a normal cell to become cancerous by inserting its own DNA into the host cell's DNA. This can happen in two ways. The virus can insert its own DNA into a host's proto-oncogene. Due to this insertion the proto-oncogene becomes an oncogene, which can then lead to the cancerous growth of the cell. Or the virus could introduce a viral genome that already contains an oncogene. The insertion of this oncogene can also lead to cancerous growth. DNA viruses implicated in human cancers are:

- **Human papillomaviruses**: sexually transmitted and linked to cervical cancer (Case study 9.17).
- **Hepatitis B and C**: linked to liver cancer;
- **Epstein–Barr virus**: a herpes virus that has been found in Burkitt lymphoma;
- **T cell leukaemia virus**: linked to T lymphocyte leukaemia;
- **Human Immunodeficiency Virus (HIV)**: linked to Kaposi sarcoma.

CASE STUDY 9.17

Cervical cancer

A number of different genes are involved in the development of cervical cancer. However, cervical cancer is strongly associated with the human papillomaviruses (HPV). Only a small number of individuals who are infected go on to develop cervical cancer.

The DNA of HPV types 16 and 18 have been found in cell lines derived from cervical cancer cells. HPV type DNA has been found integrated into both chromosomes 8q and 12q. The DNA inserted onto chromosome 8 has been found integrated into the MYC gene. This suggests that the activation of cellular oncogenes by the HPV could result in cervical cancer.

ACTIVITY 9.6

How can individuals be protected against some virally-induced cancers caused by the human papillomaviruses or Hepatitis B?

OTHER CANCER CONSIDERATIONS

Somatic or germ line?

A germ line mutation can be passed on either via the egg or sperm to the next generation. This type of mutation will then be present in every single cell of the offspring. A somatic mutation arises in a single cell within the body at any stage of life post-conception. Somatic mutations cannot be passed on to the next generation.

Familial cancers

Familial cancers are inherited through germ line mutations. Most involve the loss of function of only one copy of a tumour-suppressor gene. The loss of function of only one copy of a tumour-suppressor gene does not lead to cancer as it acts as a recessive gene. However, if an individual inherits a non-functioning copy of a tumour-suppressor gene from one parent and a functioning copy from the other parent they are at a higher risk of developing cancer as they only have one functioning gene.

Inheriting one faulty copy is known as 'one hit'. Loss of the functioning of the remaining normal gene is known as 'two hits'. The second hit is acquired through somatic mutation during a person's lifetime. It is only through two hits that the individual will develop a cancer caused by tumour-suppressor gene mutation.

Indications of familial cancer

There are certain patterns that are indicative of familial cancers.

- Several close members of the family develop the same or genetically associated cancer.
- Two individuals within the same family develop the same rare cancer.
- Early age of onset.
- Bilateral tumours in the same paired organs.
- Successive tumour development in the same individual.

Cancer is not usually inherited, however the predisposition to cancer is.

ACTIVITY 9.7

Why are most familial cancers usually associated with tumour-suppressor genes and not with oncogenes?

Cancer conditions

The genetic cause of many cancers has already been identified in humans. Active research in this field brings with it new discoveries. The amount of research being carried out is well beyond the scope of this book but a good resource for students and professionals in the field of genetics is the Online Mendelian Inheritance in Man (OMIM) website (**www. ncbi. nlm.nih.gov/omim**). The OMIM website contains relevant genetic research papers and is hosted by McKusick-Nathans Institute of Genetic Medicine, Johns Hopkins University (Baltimore, MD) and the National Centre for Biotechnology Information, National Library of Medicine (Bethesda, MD).

SUMMARY

- **Oncogenes**
  Proto-oncogenes regulate cell division and growth. Oncogenes are the mutated forms of proto-oncogenes. These genes become permanently 'switched on'. Oncogenes normally act in a dominant fashion.
- **Tumour-suppressor genes**
  Tumour-suppressor genes slow down cell division, correct DNA replication errors and instruct the cell when to die. Mutations resulting in a loss of function have to occur in both genes before cancer can develop. These genes normally act in a recessive way.
- **Environmental triggers**
  A range of chemicals, radiation and viruses have all been recognised as having a mutagenic effect on DNA.
- **Familial cancers**
  Most familial cancers are due to the inheritance of a faulty tumour-suppressor gene. Both genes need to be in a mutated form before the individual develops cancer (inheritance of one faulty copy does not result in cancer but increases the risk).

FURTHER READING

Klug, W.S. and Cummings, M.R. (2005) *Human Genetics: Concepts and Applications.* Maidenhead: McGraw-Hill

Lewis, R. (2008) *Essentials of genetics.* Harlow: Pearson Education
The textbooks above provide easy-to-read and detailed information about oncogenes and tumour-suppressor genes.

The following websites are very useful:

Macmillan Cancer Support
www.macmillan.org.uk/Cancerinformation/Causesriskfactors/Genetics/Cancergenetics/Cancergenetics.aspx
This is a British website that provides an overview of the genetics of cancer and identifies the affected genes in some of the more common cancers in the UK.

Cancer Genetics
www.cancer-genetics.org
This is also a British website that includes excellent links to some very relevant genetic sites.

Rediscovering Biology
www.learner.org/courses/biology/textbook/cancer/cancer_11.html
This is a good educational site regarding the basics of the biology of cancer. It provides good, clear explanations of the function of oncogenes/proto-oncogenes and tumour-suppressor genes.

A couple of interesting research articles regarding radiation levels while flying are:

Band, P.R., Nhu, D.L., Fang, R., Deschamps, M., Coldman, A.J., Gallagher, R.P. and Moody, J. (1996). 'Cohort study of Air Canada Pilots: cancer incidence, and leukaemia risk'. *American Journal of Epidemiology*, 145(2): 137–43

Gundestrup, M. and Storm, H.H. (1999) 'Radiation induced acute myeloid leukaemias and other cancers in commercial jet cockpit crew: A population based cohort study'. *The Lancet*, 354: 2029–31

10

GENETIC COUNSELLING

LEARNING OUTCOMES

The following topics are covered in this chapter:

- UK genetic services;
- genetic counselling;
- how to refer for genetic counselling;
- genetic testing;
- ethical issues in genetics;
- genetic data;
- internet resources.

INTRODUCTION

At the RCN congress in 2001 members agreed that nurses would act as a point of contact for patients for referral to specialist genetic services. Since then the NHS National Genetics Education and Development Centre has been established as the result of the UK government's White Paper *Our Future, Our Inheritance* (Department of Health, 2003). The National Genetics Education and Development Centre has developed workforce competencies for health professionals working in non-genetic areas of health care.

Workforce competencies

- Identify where genetics is relevant in your area of practice.
- Identify individuals with, or at risk of genetic conditions.
- Gather multigenerational family history information.
- Use multigenerational family history information to draw a pedigree.
- Recognise a mode of inheritance in a family.
- Assess genetic risk.
- Refer individuals to specialist sources of assistance in meeting their health care needs.

- Order a genetic laboratory test.
- Communicate genetic information to individuals, families and health care staff.

(National Genetics Education and Development Centre, 2007)

These competencies are for all health care professionals, although some of the competencies might not be applicable to certain people, depending on the area of work. However, all nurses, midwives and health visitors at the point of registration should be able to demonstrate the following competencies.

- Identify clients who might benefit from genetic services.
- Appreciate the importance of sensitivity in tailoring genetic information.
- Uphold the rights of all clients to informed decision making about genetics.
- Possess genetics knowledge to underpin practice.
- Understand the utility and limitations of genetic testing and information.
- Be aware of the limitations of own genetic expertise.
- Obtain and communicate information about genetics.

(Department of Health, 2003)

All health care professionals need to be able to support patients and appropriately refer on to specialised genetic services.

ACTIVITY 10.1

a. Which of these competencies can you demonstrate in practice now?

b. Which of these competencies do you think you would find it difficult to demonstrate in practice and why?

UK GENETIC SERVICES

Clinical genetic services in the UK are based in 27 regional centres. These centres offer specialist services to families that are at high risk of serious genetic disorders.They offer a range of diagnostic and counselling services.The genetic counsellors employed by the regional centres have a key role in the delivery of integrated services. Each regional centre serves a population of 2 to 6 million and provides diagnosis, risk estimation, counselling, surveillance and psychological support.The UK Department of Health (2009) has defined genetic services as:

> clinical and laboratory activity relevant to the genetic aspect of disease. The clinically-led activity, comprising clinical consultation (clinical examination and diagnosis, pedigree interpretation and investigation) as well as interpretation of genetics laboratory results, risk estimation and genetic counselling provided to patients. ... It [the service] deals with immediate and extended families, often over several generations, and provides genetic expertise for any age group affected by, or at risk of, disorders in any body system.

Each regional centre employs clinical staff (medical consultant, specialist registrar and clinical nurse specialists), scientific staff (cytogenetics and DNA laboratory staff) as well as administrative and clerical staff. In some regions there are staff that are based in district general hospitals in order to provide a locality-based service for patients.

Regional centres vary in the services offered. However, the following services are offered by all centres:

- diagnostic services;
- patient information;
- genetic counselling;
- education;
- guidance and advice.

The clinical geneticists employed in the regional centres are medical doctors who have undergone advanced training in genetics. Their role is to investigate the genetic risk factors within individuals, order appropriate tests, diagnose and counsel. They also provide advice, education and training to other health care professionals, as well as undertaking research in clinical, biomedical, psychological and service-related issues.

Most of the genetic nurse specialists within the regional centres are employed as genetic counsellors. They provide a non-directive provision of information to enable the patient to make open and free choices regarding possible testing. They also provide psychological support both before and after diagnostic testing. Genetic nurses are registered as genetic counsellors with the Association of Genetic Nurses and Counsellors (AGNC) Registration Board and are bound by their Code of Ethics.

A patient's initial contact with the genetic services might be with a genetic counsellor or with the clinical geneticist. If the initial contact is via the genetic counsellor then this might take place in the genetics clinic or in the patient's own home.

ACTIVITY 10.2

Find out where your nearest regional centre is by visiting The British Society for Human Genetics website. **www.bshg.org.uk/genetic_centres/uk_genetic_centres.htm**

GENETIC COUNSELLING

The EuroGentest Network of Excellence (2008) defines genetic counselling as a process of communication that deals with the occurrence, or risk of occurrence, of a possible genetic disorder. The aim of the counselling session is to enable the consultand to:

- understand the medical facts of the disorder;
- appreciate how heredity contributes to the disorder and the risk of reoccurrence in family members;

- understand the options for dealing with the risk of occurrence;
- use the genetic information in order to promote health, minimise psychological stress and increase personal control;
- choose a course of action that is appropriate to them;
- make the best possible adjustment.

Genetic advice to patients and their families is given in a non-directive, non-judgemental and supportive manner. Non-directive counselling means giving the individual enough information so that they can make a free and informed choice. Supporting people to make their own choices is the main focus of the genetic counsellor's work. There are some circumstances where using a non-directive approach is not always possible such as, for example, if a couple wanted to plan the sex of a child as they already had three daughters and desperately wanted a son. In a case like this it may be impossible for the counsellor to remain neutral.

The initial contact with a genetic counsellor will involve the gathering of personal and family medical history. At the end of the session, a plan of care may be made regarding further information-gathering sessions, exploration of options or testing. Most patients will attend a minimum of two genetic counselling sessions, one pre-test and one post-test. However, it is not unusual for an individual to attend multiple counselling sessions.

In many cases a patient has received a diagnosis before being referred on for genetic counselling but, in some instances, supportive counselling is needed before a definite diagnosis. Most genetic counselling sessions are based around the gathering of family history where culture and ethnicity are also taken into account. This is done to aid risk assessment as some genetic disorders are more common in some populations (see Table 10.1).

It is also important for the counsellor to assess what the consultand wants to know. Some individuals might prefer to have regular medical surveillance for a pre-symptomatic condition rather than a definite test result. Family history and empiric risk calculation can help the individual to make a choice towards testing or not. Guidelines covering predictive testing in children for late onset disorders, familial cancers, and for the interpretation and reporting of genetic tests, have been developed and are used in the regional centres.

Individuals generally seek genetic counselling for one of two main reasons: pre-natal diagnosis or a genetic disorder with the family.

Pre-natal diagnosis

Pre-natal genetic counselling usually involves informing the prospective parents about the familial and empiric risks, giving an explanation of the tests available and examining whether the benefits of the test outweigh the risks of the condition.

Table 10.1 *High-frequency alleles in certain populations*

Population	Disease
Afrikaaners	Porphyria variegata Familial hypercholesterolaemia Huntington's disease
Amish	Chondroectodermal Dysplasia Ellis–van Creveld syndrome
Ashkenazi Jews	Tay–Sachs disease Gaucher disease Canavan disease Breast cancer (BRCA1 mutations)
Central and South East Asian, Southern European, Middle Eastern and North African	Beta-Thalassaemia
Chinese, South East Asian, Southern European, Middle Eastern, Indian, African, Pacific Islands	Alpha-Thalassaemia
Finns	Congenital nephrotic syndrome
Irish	PKU
Lebanese, French Canadians	Familial hypercholesterolaemia
Northern Europeans	Cystic Fibrosis Hereditary haemochromatosis
Ryukyan Islands (Japan)	Spinal muscular atrophy
West Africans	Sickle cell anaemia

Familial disorder

For a recessive disorder, the affected individual is usually a child. Parents may want to know the risks of having a second child with the same disorder, or may want a definite diagnosis for the affected child. Counselling individuals for adult onset disorders presents as a challenge for some individuals as to whether they want to know if they will develop the disorder in the future. The counsellor will ensure that the individual is aware that having a positive test result for an affected gene does not always mean that the individual will develop that disorder. It is important that there is understanding that a 'diagnosis' of a condition is made on the presence of symptoms and not the presence of an affected gene.

ACTIVITY 10.3

Describe in your own words what a genetic counsellor does.

BECOMING A GENETIC COUNSELLOR

Back in 2001, the Association of Genetic Nurses and Counsellors (AGNC) formed a registration board for genetic counsellors practising in the UK and Eire. This board manages the professional register of practitioners and sets the standards for entry into the profession as well as advising and investigating concerns regarding fitness to practise. All registrants are bound by the AGNC Code of Ethics (see below).

Box 10.1 (AGNC Code of Ethics)

A. Self-awareness and development

Genetic counsellors should:

- be aware of their own physical and emotional health and take appropriate action to prevent an adverse impact on their professional performance;
- report to an appropriate person or authority any conscientious objection that may be relevant to their professional practice;
- maintain and improve their own professional education and competence.

B. Relationships with clients

Genetic counsellors should:

- enable clients to make informed independent decisions, free from coercion;
- respect the client's personal beliefs and their rights to make their own decisions;
- respect clients, irrespective of their ethnic origin, sexual orientation, religious beliefs and gender;
- avoid any abuse of their professional relationship with clients;
- protect all confidential information concerning clients obtained in the course of professional practice: disclosures of such information should only be made with the client's consent, unless disclosure can be justified because of a significant risk to others;
- report to an appropriate person or authority any circumstance, action or individual that may jeopardise client care, or their health and safety;
- seek all relevant information required for any given client situation;
- refer clients to other competent professionals if they have needs outside the professional expertise of the genetic counsellor.

C. Relationships with colleagues

Genetic counsellors should:

- collaborate and co-operate with colleagues in order to provide the highest quality of service to the client;

- foster relationships with other members of the clinical genetics team, to ensure that clients benefit from a multidisciplinary approach to care;
- assist colleagues to develop their knowledge of clinical genetics and genetic counselling;
- report to an appropriate person or authority any circumstances or action which might jeopardise the health and safety of a colleague.

D. Responsibilities within the wider society

Genetic counsellors should:

- provide reliable and expert information to the general public;
- adhere to the laws and regulations of society. However, when such laws are in conflict with the principles of practice, genetic counsellors should work towards change that will benefit the public interest;
- seek to influence policy makers on human genetic issues, both as an individual and/or through membership of professional bodies.

Under the AGNC registration system, practitioners wishing to train as genetic counsellors must satisfy the registration board that they have one of the following set of entry level qualifications:

a. Graduate with professional qualification and experience:

- relevant First or Master's degree (in nursing, biological or social sciences);
- appropriate professional qualification (in nursing, midwifery or social work) and maintenance of current registration;
- minimum of two years' professional clinical experience in a health or social care setting;
- basic counselling training of at least 120 guided learning hours.

or

b. Master of Science Degree in Genetic Counselling:

- substantial clinical and experiential component;
- completion of basic training in counselling skills of at least 120 guided learning hours.

(AGNC, 2006)

The Genetic Counselling Registration Board has approved two MSc in Genetic Counselling courses in the UK, one at Manchester University and one at Cardiff University.

The necessary qualifications and work experience prior to entry into the profession is a minimum of five years for both the professional level entry and for the MSc level entry. Registration as a genetic counsellor can only be undertaken after a training period of two years following entry into the profession. During the two-year training period the practitioner has to demonstrate competencies by the submission of a portfolio of evidence while working in an area of genetic health care. By the time of registration, genetic counsellors will have completed at least seven years of training.

MAKING AN APPROPRIATE REFERRAL

Nurses, midwives and health visitors are not expected to diagnose genetic conditions but are expected to be able to identify individuals who might benefit from specialist genetic services and to identify those at risk. Appropriate referrals to the specialist genetic centres include:

1. **Pre-natal referrals**

- Pre-pregnancy counselling where there is a known family history of a condition.
- Women who have experienced multiple pregnancy losses and/or neonatal deaths.
- Parents who have a child with a genetic disorder and are seeking a pre-natal diagnosis.
- Pregnant women who have abnormal pre-natal results.
- Couples who are related and are planning a family (consanguineous mating).

2. **Developmental referrals**

- Parents of a child who has a genetic disorder detected by routine newborn screening.
- Parents of a child with learning difficulties or a developmental delay, referred to see if a diagnosis can be made.
- Parents of a child who has a known genetic disorder referred for counselling.
- Parents of a child who has dysmorphic features.

3. **Family history referrals**

- Adults or children with a known genetic condition referred for counselling.
- An individual with a possible genetic condition wanting specialist advice regarding the condition.
- A person with a family history of a possible genetic condition, to establish the risks and their options.
- A person with a personal or strong family history of breast/ovarian cancer or bowel cancer.
- An individual with one or more family members with a birth defect.
- An individual who has one or more family members with an adult onset health condition such as cardiovascular disease, diabetes or dementia occurring at an early age (usually before the age of 60 years).

National Service Frameworks

The management of genetic disorders and indications for referral to genetic services has been addressed in several National Service Frameworks. These frameworks highlight the importance of recognising individuals with, or at risk of developing, genetic conditions, making appropriate referrals to specialist genetics centres and the importance of health education and health promotion.

- **The Coronary Heart Disease National Service Framework (2000)**: standards 3 and 4 highlight the importance of preventing coronary heart disease in high-risk groups and the importance of awareness of inherited conditions.
- **The National Service Framework for Long Term Conditions (2005)**: recognises the importance of referring individuals to a genetics centre if they have a condition that is genetic in origin.
- **The National Service Framework for Diabetes (2001)**: recognises that maturity onset diabetes can also affect the young due to genetic disorders of insulin metabolism.
- **The National Service Framework for Renal Services (2004)**: recommends monitoring individuals who have a family history or genetic risk of kidney disease.
- **The National Service Framework for Children, Young People and Maternity Services (2004)**: identifies individuals who have a family history of a genetic disorder or who are concerned about a familial disease/disorder as individuals who need specialist pre-conception advice and support.

Referring patients with cancer

Around 25 per cent to 35 per cent of some common cancers, which includes breast, ovarian, colorectal and prostate cancers, have a high heritable component. Following the publication of the *Our Inheritance, Our Future* White Paper (Department of Health, 2003), the UK government has funded cancer genetic services. Most oncology centres now have specialist clinical genetics services provided in partnership with regional genetic centres. Oncology patients and those at high risk of developing cancer have access to cancer genetic counsellors based at the oncology centres.

The minority of cancer cases follow Mendelian inheritance patterns and specific genes have been identified in some cancers. The management of such cancers is recognised in guidelines produced by the National Institute for Health and Clinical Excellence (NICE) and the Scottish Intercollegiate Guidelines Network (SIGN). Most of the UK genetic regional centres and oncology genetic centres have developed their own referral criteria based on the NICE and SIGN guidelines.

Referral criteria for familial cancers are based on a tripartite system.

- **'Low risk'** – the patient is encouraged to undertake normal screening (the same as for the general population).
- **'Medium risk'** – the patient is encouraged to have regular screening.
- **'High risk'** – the patient should be referred for genetic counselling and pre-symptomatic testing is made available.

High-risk individuals who can be referred to specialist cancer genetics centres are any of the following (see Figures 10.1 to 10.4).

BREAST CANCER	
Any of the following indicates an appropriate referral	
	Female breast cancers only
	One 1st degree relative diagnosed before the age of 40
+	**One** 1st degree relative **and one** 2nd degree relative diagnosed before the age of 50
	Two 1st degree relatives diagnosed before the average age of 50
	Three or more 1st or 2nd degree relatives diagnosed at any age
	Male breast cancer
	One 1st degree relative diagnosed at any age
	Bilateral breast cancer
	One 1st degree relative where 1st primary diagnosed before the age of 50 *For bilateral breast cancer, each breast has the same count value as one relative.*

Figure 10.1 *Breast cancer*
Source: NICE (2006).

BREAST AND OVARIAN CANCER	
The following indicates an appropriate referral	
+	**One** 1st or 2nd degree relative with ovarian cancer at any age **and one** 1st or 2nd degree relative with breast cancer at any age (one should be a 1st degree relative)

Figure 10.2 *Breast and ovarian cancer*
Source: NICE (2006).

COLORECTAL CANCER	
The following indicates an appropriate referral	
	One 1st degree relative diagnosed before the age of 45
	Two first degree relatives, diagnosed at any age.
+	**One** first degree **and one** 2nd degree relative on the same side of the family, diagnosed at any age.
+	**One** first degree relative **and two** other relatives from the same side of the family, diagnosed at any age.

Figure 10.3 *Colorectal cancer*
Source: NICE (2004).

OTHER CANCERS

The following indicates an appropriate referral

- A family history of a single gene cancer (e.g. multiple endocrine neoplasia, retinoblastoma)
- Multiple cancers in **one** individual
- **Three or more** relatives with any cancer at an earlier age than expected in the general population.
- A family history of **two or more** cases of the same cancer in 1st or 2nd degree relatives.
- **Three or more** relatives with cancers of breast, colorectal, ovary, prostate, pancreas, melanoma, thyroid or other non-melanoma skin tumours or sarcoma

Figure 10.4 *Other cancers*
Source: NICE

Box 10.2 Information to remember when taking a family history

All relatives must be on same side of family and be blood relatives of the consultee and of each other.

First-degree relatives: mother, father, daughter, son, sister and brother

Second-degree relatives: grandparent, grandchild, aunt, uncle, niece and nephew; half sister and half brother

Third-degree relatives: great grandparent, great grandchild, great aunt, great uncle, first cousin, grand nephew and grand niece

NICE (2006)

Any condition that affects several members of a family, possibly at a young age (for example, hypercholesterolaemia, cardiovascular disease, diabetes) increases the risk that the condition might be inherited. In order to determine if the condition is a familial one, the health care professional needs to be able to construct a pedigree chart of the family history (Box 10.2). This needs to be done in a sensitive manner, especially when asking about bereavements, miscarriages, infertility and adoptions.

How to refer a patient for genetic counselling

About 50 per cent of all clinical referrals for genetic counselling originate from primary care and 50 per cent from secondary care. Although local arrangements differ, most regional centres have well-established links with obstetric units, paediatric units, oncology units and adult specialities.

How referrals are made varies between different regional centres. Referral letters can be addressed to a specific person or to the genetics team. Some centres accept referrals directly from patients, although most referrals are sent via a doctor, nurse, midwife or health visitor. The requirements for a referral for all centres include:

- patient information – including name, address, post code, contact number, date of birth and NHS number;
- details of the referrer;
- details of the GP (if not the referrer);
- information about the diagnosis if known;
- family history;
- for cancer referrals – age at diagnosis and primary site of tumour in all family members.

As with all documentation, the information should be contemporaneous, dated, factual and non-judgemental (NMC, 2009).

GENETIC COUNSELLING FOR TESTS

Pre-test counselling

If one of the options discussed during a consultation is the availability of a genetic test, then the counsellor's role is to inform the individual about the purpose of that test. Issues such as the psychological impact of the test results and consequences to the individual and other family members are also considered. Pre-test counselling includes discussions regarding the need to inform relatives about the test result, especially with conditions where early diagnosis may improve health.

Post-test counselling

The initial focus of a genetic counselling session following disclosure of test results is on the emotional impact on the patient. Implications of the test results to the individual and their family are discussed and follow-up sessions may be arranged. A strategy to inform relatives may have to be agreed upon, although this is sometimes more appropriately discussed in a follow-up session, depending on the individual case. Written material may be offered so that the individual can hand the information out to relevant family members.

Genetic tests offered

Different forms of genetic tests can be used to detect a gene mutation, linked haplotype or a chromosomal alteration. A genetic test will not yield results of all mutations as the whole genome is not examined. An individual being tested for a particular genetic disorder will not be examined for other conditions and therefore is not told of any other non-related mutated genes that might increase their risk of developing other conditions. Genetic tests can only be used for medical reasons as the Convention of the Council of Europe on Biomedical Issues prohibits genetic testing for any other reason.

Diagnostic tests

These are performed in order to confirm the presence of a genetic condition in a symptomatic individual. This test is not very different for the individual to other medical tests that are performed to confirm a diagnosis. However, a positive result in a diagnostic test does have implications for family members.

Predictive tests

Predictive tests are usually performed on individuals who are at high risk of developing a specific genetic disorder. This type of test does not confirm whether the individual will develop the condition, only confirming the risk.

Pre-symptomatic tests

These are similar molecular tests to predictive tests. The only difference is that pre-symptomatic tests are used for adult onset conditions that show 100 per cent penetrance. A positive result indicates that the individual will develop the condition.

Susceptibility tests

This is also known as risk profiling. This test is used to detect the presence of genetic markers that are associated with an increased risk of developing certain genetic multifactorial disorders. The clinical validity and reliability of these tests is questionable.

Pharmacogenetic tests

These are used to detect genetic susceptibility to adverse drug reactions or reduced efficiency of some medications.

Carrier testing

This is done to detect the presence of a recessive disorder-causing allele. The individual is not affected but has a 50 per cent chance of passing the affected allele on to the next generation.

Pre-natal tests

Pre-natal tests are performed during pregnancy on women who are considered at high risk of having a child with a severe genetic disorder.

Pre-implantation test

This test detects mutations or chromosome alterations in one or two embryo cells pre-IVF. It is only performed on embryos where there is a known genetic disorder (or that are at a high risk of a genetic disorder).

Genetic screening

This test is systematically offered to a part of the population that is considered to be at risk of a genetic disorder. Genetic screening does not usually involve genetic counselling unless a genetic alteration has been detected. Health care professionals who offer genetic screening should inform patients about the condition being screened for, the individual testing methods, implications of the results, and ensure the individual has the freedom of choice regarding participation.

ACTIVITY 10.4

Do you think genetic testing should be performed when there is no treatment available?

ETHICAL ISSUES

The four main principles used to guide ethical practice in health care are autonomy, beneficence, non-maleficence and justice (Beauchamp and Childress, 2001).

Autonomy

Autonomy is the belief that a person has the right to make their own decisions in life, provided that the consequence does not harm or violate another person (Hawley, 2007). An individual is free to refuse any genetic test. Everyone has a right to know as well as the right not to know. In order for an individual to make autonomous decisions regarding their health they need to be fully informed and free from external controls. Genetic counselling is delivered in a non-directive fashion, providing education and options for the patient to make their own decisions regarding any possible testing.

Beneficence

Beneficence is the duty to do good. Care is aimed towards the wellbeing of the patient (Hawley, 2007). After providing information regarding the genetic risks, the available tests and options available, the patient's choice should be supported and their needs and preferences respected.

Non-maleficence

This means to do no harm. There are many situations in genetic health care where harm could occur, such as, for example, unconsented disclosure of genetic information to a third party or giving the patient incorrect genetic information. Professionals should always recognise their own limitations and report to, or seek relevant advice and information from, other relevant professionals.

Justice

Justice ensures that there is equity of care. This implies that all individuals should have access to available services. This is not always the case in the health service due to the financial constraints within the NHS. Many of the UK regional centres limit the availability of some genetic tests. These are usually tests that have no influence on health outcomes (the patient's condition may have already been diagnosed through other medical tests). The allocation of resources is carefully managed throughout the NHS and genetic services are no exception.

Confidentiality and consent

The guiding tenets of ethical care in practice are confidentiality and consent.

Confidentiality

Genetic information, like any other information, should be protected by the principle of confidentiality (Case study 10.1). Confidentiality is a fundamental part of professional practice that protects the individual's human rights (European Convention of Human Rights, 1998). Disclosure of information given in confidence should only be made with the patient's consent.

There are some exceptions to this where a health professional is allowed to breach confidentiality under certain circumstances. In situations where there is a possibility of harm to a third party (or to the patient themselves) it is allowable to breach confidentiality. Breaching confidentiality is also allowable if public interest in breaching outweighs public interest in maintaining confidentiality. Disclosure without permission is only permitted if there are reasonable grounds to protect the patient, family or the community.

CASE STUDY 10.1

The Genethics Club is a national forum for the discussion of practical ethical problems encountered in genetic centres within the UK. Anyone working within a clinical genetic department can become a member of this club that meets up three times a year. The following two case studies are examples of confidentiality issues that have been discussed at the club. What are your views on these case studies?

Case 1

Jim (60 years old) has been diagnosed with Huntington's disease, confirmed through genetic testing. There is no known family history of Huntington's disease but several deceased relatives are thought to have had Alzheimer's disease. Jim himself was initially diagnosed with Alzheimer's disease in his early 50s, but the diagnosis was reviewed following brain scans and revised to Huntington's disease at the age of 57. Jim and his wife Mary are adamant that they do not want any of their four children (aged 30 to 37) to be informed of the diagnosis, and they would rather 'they continued to think it was Alzheimer's'. Jim and Mary have been offered help in communicating the diagnosis to their children but they have declined. Recently, one of their daughters was referred to the genetic team to discuss her family history of Alzheimer's disease. What should the genetic team say to her during the appointment?

Case 2

Polly and Richard, a married couple, have been referred to a clinical genetics department following the diagnosis of a rare autosomal recessive condition in their new-born baby. The disorder is severe and debilitating and there is a high chance that the child will die in the first year. During their first session with a genetic counsellor they are informed that there is a 25 per cent chance that a future baby will have the same disorder. Following their first appointment, Polly rings the genetic counsellor and informs them that Richard is not the father of the child, but he is not to be informed of this. As the mutation is rare in the population, the risk to a future child is in fact negligible. At the next counselling session, Richard does most of the talking and wants to explore options of having further children by artificial insemination by donor. What should the counsellor do in this situation?

If Richard and Polly were to divorce at some point in the future and Richard wanted to have a child with another woman, should Richard be informed of his low risk of having an affected gene?

In the genetics health care setting, the genetic information gained from one individual may have implications for other members of the same family. Permission to share that information within the family must be sought. Most patients will readily give consent for information to be disclosed. However, in circumstances where the patient refuses to allow disclosure of relevant information, the Nuffield Council on Bioethics (1993) has outlined that an individual's desire for confidentiality can be overridden in exceptional circumstances where family members may be at risk from the same genetic condition. This usually only applies to genetic conditions where medical intervention or prevention that can improve health is available. The General Medical Council issued guidelines for doctors in 2009 that allow doctors to breach genetic confidentiality. This guidance clarifies that doctors should explain to the patient if their family might be at risk of inheriting a genetic condition, and that it is the doctor's responsibility to protect those who may be at risk. Doctors can disclose genetic information to relatives, even when patients object, if there are compelling reasons for doing so.

ACTIVITY 10.5

When is it appropriate to breach confidentiality?

Consent

To ensure that patients understand the risks and benefits of health care choices, informed consent is vital in the decision-making process. Under the Human Tissue Act of 2005, it is an offence to obtain DNA for the purpose of analysis without consent. For individuals considering genetic testing the following points are discussed with a genetic counsellor before an informed consent can be obtained:

- the voluntary nature of testing;
- the risks, limitations and benefits of testing/not testing;
- the alternatives to testing;
- details of the testing process;
- the possibility that the test may reveal unexpected results;
- the privacy and confidentiality of the results;
- information may be shared with other health professionals;
- the use and sharing of information with other family members for their health benefit;
- the storing of samples (samples may be used for quality assurance, education and training);
- potential consequences relating to the results (impact on health, possible psychological effects, treatment/prevention options).

Once a test result is known it cannot be unknown.

Recording consent

The Joint Committee on Medical Genetics (2006) has recommended that consent should be obtained in a written format and they have issued a consent document framework for the use of the UK regional centres (see Figure 10.5).

Name:

Date of birth:

Address:

During this consultation we have discussed the following issues, and you have agreed to the uses shown below. Please cross out the words 'Discussed', 'Agreed', 'Not applicable' as appropriate for each use.

1	I agree to analysis of the sample for	Discussed	Agreed	Not applicable
2	I agree to the sample being stored in case future checks or tests are needed	Discussed	Agreed	Not applicable
3	I would like to be contacted before further tests are done on the stored sample if new tests become available	Discussed	Agreed	Not applicable
	OR			
4	I am happy for further diagnostic tests on the stored sample to be undertaken without being contacted	Discussed	Agreed	Not applicable
5	I agree that information and test results may be shared to help other family members	Discussed	Agreed	Not applicable

Signed.. Date...
(Clinician)

Please confirm your agreement by adding your signature to this form below:

Signed..
(Patient/Parent)

We may keep any leftover samples to check the quality of our results for other patients. We make sure that nobody knows whose sample is helping us to do this.

Copy: Records
Patient/Parent

Figure 10.5 *Record of consultation and agreement over genetic testing and sharing of information*
Source: Joint Committee on Medical Genetics (2006)

ACTIVITY 10.6

Who do you think should have access to an individual's genetic information?

Who can give consent?

- **Adults** – all adults can consent if they are mentally competent to do so.
- **Children** – parents may give consent for genetic testing in children. However, there are additional ethical issues that need to be considered when children are involved. Testing is usually only indicated in order to make a diagnosis where treatment or surveillance is available for that particular genetic condition. Testing for adult onset disorders or carrier testing should generally be delayed until the child is old enough to make their own informed choice of whether to be tested or not.
- **Adults with mental incapacity** – an adult who is lacking in mental capacity is unable to give consent for genetic testing. However, a genetic test can be ordered by their doctor if it is viewed as in the best interest of the adult concerned (Mental Capacity Act 2005; Adults with Incapacity Act 2000). Testing is only done if it is deemed necessary for the treatment or care of the individual concerned. The Joint Committee for Medical Genetics (2006) has recommended that before carrying out a genetic test on an individual with mental incapacity, the possibility of other family members (who are capable of consenting) providing the relevant genetic information should be explored first.

ACTIVITY 10.7

Should parents have the right to have their children tested for adult onset disorders?

The use of genetic data

Genetic data is usually generated through DNA testing in nationally approved clinics. This data can include:

- tests for the presence or absence of an allele;
- tests for the DNA sequences near the gene (genetic markers);
- tests for gene products or proteins.

Genetic information can be derived from an individual's genotype or from family members. In clinical practice, genetic data is used to either predict or confirm a diagnosis. The regulation of this data varies from country to country. The UK regional centres retain the DNA samples from an estimated 1.5 million UK citizens and these samples are held indefinitely. The data is often shared between some regional centres due to the specialities of expertise in some centres. This is likely to change as Connecting for Health and other medical records systems are established through the National Programme for Information Technology in the NHS. These new systems will enable more information sharing within the NHS. The risk in genetic services is not any different from other NHS departments, but security mechanisms will have to be established to reduce the risk of access to family health data.

Genetic data is also held outside the NHS. In the UK, the publicly run Biobank database holds DNA samples from around half a million UK citizens. The Biobank is funded by the Medical Research Council and the Wellcome Trust and has government approval. The purpose of the Biobank is to collect large numbers of DNA samples to aid research. Many mutations are rare and the relevance of these mutations in certain genetic conditions can only be established through large-scale studies. Genetic research is carried out on Biobank samples as well as in UK regional centres. The Biobank is a research centre and does not provide medical advice to individuals.

Other countries have different research establishments, some of which offer postal tests to the public. Ethical problems can arise regarding who owns the genetic data in these commercial testing companies. In 1998 the Icelandic parliament allowed the whole population's individual health records to be passed onto a large database without patients' explicit prior consent (individuals were given the opportunity to opt out). The DNA data of the Icelandic population was in the hands of the privately owned DeCode Company. This company held the DNA profiles belonging to most Icelandic people and thousands of other people who had paid for postal tests. The DeCode Company's policy was not to disclose information to third parties such as insurers, employers and doctors. The company went into administration at the end of 2009 and was taken over by a new company, Saga Investments, in 2010. This new company planned to sell on some of the data to pharmaceutical companies and academic research establishments. Although individual confidentiality will still be maintained it does raise the fact that individuals who buy 'over the counter tests' cannot be certain that their data cannot be put to other uses.

ACTIVITY 10.8

Who 'owns' the genetic information encoded in an individual's DNA?

ACTIVITY 10.9

Some commercial companies that provide genetic tests directly to the public use 'virtual genetic counselling'. This is interactive computer software that provides basic information and a calculated risk score for developing the genetic disorder in question. What are the advantages and disadvantages of using a 'virtual counsellor'?

Non-medical uses of genetic data

Insurance

The use of genetic data for life assurance has not been debated as much in the UK and Europe as it has been in the USA. Currently, legislation prevents insurance companies using genetic information in Austria, Denmark and Norway. However, despite the lack of legislation in The Netherlands, France, Sweden and the UK there is a moratorium in place. In the UK a moratorium exists where no insurer would require the disclosure of a genetic test result in order to determine the availability or terms of insurance. This moratorium was renewed in the UK in 2011, and the agreement has been extended to 2017. The European Council Convention on Human Rights and Biomedicine now places a ban on all forms of gene-based discrimination.

Paternity testing

Paternity testing involves the DNA analysis of the child, its mother and that of the presumed father. It should only be carried out in the best interest of the child. As paternity testing is considered as non-medical genetic testing, the tests are not available on the NHS, even if requested by a court. Private paternity tests vary in price, with many commercial companies offering this service.

Forensic testing

DNA profiling for solving criminal offences is increasingly being used by the police. In the UK a separate genetic database is used by forensic services. The UK National Criminal Intelligence DNA Database was set up in 1995 and now contains samples from over 5 million individuals. This database is run by the Forensic Science Service, which is under contract to the Home Office. The Criminal Justice Act 2003 allows DNA to be retained from individuals who have been arrested by the police. The genetic data held on this database is owned by the police authority who initially took the sample and the data is kept indefinitely. This practice differs in Scotland and in the Isle of Man, where only genetic data from convicted criminals is included in the database. The UK National Criminal Intelligence DNA Database has no links with any other databases. Police have no automatic right to access the NHS genetic database or to medical records.

ACTIVITY 10.10

Do you think that the police should have access to the NHS genetic database? Explain your answer.

INTERNET RESOURCES FOR GENETIC EDUCATION

Health care professionals should be able to guide patients to factual and accurate educational resources to enable the patient to become empowered and 'expert' (Muir Gray, 2002). Professionals should also be aware of where to get further information for the benefit of their own practice. There is now a great wealth of information available to both patients and professional practitioners on the web. This can present as both a blessing and as a problem in that information can be quickly and easily obtained but might prove distressing and confusing for some.

Genetic internet databases have been in existence for quite a while and contain information and resources on specific genetic disorders and genetic syndromes. The following examples present detailed information regarding the individual genes that are linked to specific disorders. These sites are useful for individuals who already have some knowledge of what they are looking for.

- **Genetic Alliance** – provides information on a range of different genetic conditions (and also has a link to the European Alliance) **www.geneticalliance.org**
- **GeneReviews** – provides information on various genetic conditions and different genetic tests. **www.ncbi.nlm.nih.gov/sites/GeneTests/review**
- **National Organization for Rare Disorders (NORD)** – this is a database of rare disorders. **www.rarediseases.org**
- **Online Mendelian Inheritance in Man (OMIM)** – contains a wealth of information on different genetic disorders, gene mutations and research papers. The site also provides access to Medline. **http://ncbi. nlm.nih.gov/omim**

There are numerous other sites that are considered excellent for providing an overview of a specific genetic disorder and can be used as a starting point for information gathering. Many of these sites are useful for patients and their families.

- **Centre for Genetics Education (Australia)** – provides fact sheets for different genetic disorders. **www.genetics.edu.au**
- **Genetics Home Reference** – provides information on the science of genetics and information on different genetic conditions. **www.ghr. nlm.nih.gov**
- **National Human Genome Research Institute** – contains factsheets for some genetic disorders but also has links to other relevant sites. **www.genome.gov/10001204**
- **Patient UK** – this website contains useful information for patients on many different genetic disorders. **www.patient.co.uk**

Lay societies also provide a lot of useful information for patients, their families and for the health care professional. There are hundreds of different lay societies, which are too numerous to mention here. However, many of these societies have combined to form umbrella groups that can be accessed via the following websites.

- **CLIMB** (Children living with inherited metabolic diseases) – **www. climb.org.uk**
- **Contact a Family** – provides information and support to parents of children who have a genetic disorder. **www.cafamily.org.uk**
- **Genetic Alliance UK** – this website lists contact details of most UK support groups. **www.geneticalliance.org.uk**

As with other areas of health care, there has been a rapid increase in the number of scholarly journals available in genetics. Web searches through OVID, PubMed or Google Scholar will provide abstracts of most articles, with many articles available in full. The internet provides a wealth of information but any information given to a patient has to be evaluated as relevant, accurate and up to date (NMC, 2008).

ACTIVITY 10.11

Alice is 16 weeks pregnant and is being treated in hospital for a deep venous thrombosis. She has been informed by the doctor that the thrombosis has occurred as a result of a genetic condition called paroxysmal nocturnal haemoglobinuria. Alice confides in you that she did not understand much of what the doctor told her but, as the condition is genetic, she is worried for her unborn child. She asks you if the baby is likely to have this condition as well.

By searching through relevant internet resources, what information can you give to Alice regarding the risk of inheritance to her unborn child?

Genetics is a rapidly growing field within health care and all health professionals need to keep up to date with the latest research evidence and advances in genetics so, that this can be applied in practice for the benefit of patients.

SUMMARY

- All health care professionals have to be competent in recognising the relevance of genetics in health care practice, have the ability to inform, educate and support patients and identify individuals who might benefit from specialist genetic services.
- Specialist genetic services in the UK are coordinated from regional genetic centres and provide clinical consultations, genetic counselling, risk estimations, genetic testing and diagnosis.
- Appropriate referrals include any individuals who might have a genetic condition, are a possible carrier of an affected gene, or who might be at risk of developing a genetic disorder.
- Cancer genetic services accept referrals from patients who are considered to be at 'high risk' of cancer. High risk classifications have been established through NICE guidelines.
- All patient information is confidential. Disclosure of confidential information without consent can only occur in exceptional circumstances where the harm from keeping confidentiality outweighs the harm from breaching it.
- Consent for testing has to be informed and free from any external pressures. Parents are able to consent on behalf of children but this is not recommended for most adult onset disorders.

- Within the NHS, all genetic data on patients is generated in nationally approved clinics. Different genetic databases also exist outside the NHS in the form of research establishments, commercial companies and the Criminal Intelligence DNA Database.
- Health care professionals need to be able to communicate relevant genetic information to patients and to other professionals. All information used should be accurate, peer-reviewed and up to date. Internet resources provide valuable information for both patients and health care professionals.

REFERENCES AND FURTHER READING

Adults with Incapacity (Scotland) Act 2000. Edinburgh: Scottish Government

Beauchamp, T.L. and Childress, J.K. (2001) *Principles of biomedical ethics.* Oxford: Oxford University Press

Department of Health (2003) *Our Inheritance, our Future: Realising the potential of genetics in the NHS.* London: Department of Health

Department of Health (2009) *Genetic service definition number 20.* **www.specialisedservices.nhs.uk/doc/medical-genetic-services-all-ages**

EuroGentest Network of Excellence (2008) *Harmonizing genetic testing across Europe. Genetic Counselling Definition.* **www.eurogentest.org**

General Medical Council (GMC) (2009) *Confidentiality guidelines.* London: GMC

Genetic Interest Group (1998) *Confidentiality guidelines.* London: Genetic Interest Group

Hawley, G. (2007) *Ethics in clinical practice: An interprofessional approach.* Harlow: Pearson Education

Human Tissue Act 2005. London: HMSO

Joint Committee on Medical Genetics (2006) *Consent and confidentiality in genetic practice: Guidance on genetic testing and sharing genetic information.* London: Genetic Interest Group

Mental Capacity Act 2005. London: HMSO

Muir Gray, J.A. (2002) *The resourceful patient.* Oxford: Rosetta Press

National Genetics Education and Development Centre (2007) *Enhancing patient care by integrating genetics in clinical practice: UK workforce competencies for genetics in clinical practice for non-genetics healthcare staff.* **www.geneticseducation.nhs.uk**

National Institute of Clinical Excellence (NICE) (2004) *Guidance on cancer services. Improving outcomes in colorectal cancers.* London: Department of Health

National Institute of Clinical Excellence (NICE) (2006) *Familial breast cancer. The classification and care of women at risk of familial breast cancer in primary, secondary and tertiary care.* London: Department of Health

Nursing and Midwifery Council (2008) *Code of professional conduct.* London: NMC

Nursing and Midwifery Council (NMC) (2009) *Record keeping: Guidance for nurses and midwives.* London: NMC

The Criminal Justice Act 2003. London: HMSO

ANSWERS TO THE ACTIVITIES

Chapter 1

Activity 1.1:
- Haploid is 23
- Diploid is 46

Activity 1.2:
Crossing-over of chromosomal material during meiosis allows for the exchange of DNA from maternally inherited chromosomes and paternally inherited chromosomes. The resulting daughter cells are therefore not identical to the original cell. Mitosis does not normally involve any 'crossing over' of genetic material.

Activity 1.3:
Major phases include interphase and mitosis. Interphase includes phases G1, where chromosomes are replicated and G2, where the cell recovers from cell division and develops optimum cell size and function. Mitosis is the division of the cell and includes the phases prophase, metaphase, anaphase and telophase.

Activity 1.4:
Germ cells contain half the chromosomal complement of a normal cell in order to prevent doubling of chromosome numbers in each generation.

Activity 1.5:

Base sequence on mRNA	CCU CAA AGU GGU GUU CGA
Base sequence on DNA	GGA GTT TCA CCA CAA GCT

CCU: Proline, CAA: Glutamine, AGU: Serine, GGU: Glycine, GUU: Valine, CGA: Arginine.

Activity 1.6:
20 amino acids

Activity 1.7:
Chromosomes are structures that are made up of DNA and protein molecules. The genes are part of the chromosomal structure. Genes are sequences of nucleotides that produce proteins through transcription and translation. Genes are the basic units of heredity.

Activity 1.8:
- Chromatin: coiled up DNA and protein molecules that store genetic information.
- Nucleus: chromosomes are stored in the nucleus; genes are expressed here and are copied by RNA, which migrates out into the cytoplasm.
- Ribosome: an organelle that functions in protein synthesis. Both transfer RNA and messenger RNA gather at the ribosome for protein formation. Transmission of genetic material.
- Mitochondrion: has its own genome. Functions in the storage and expression of genes.
- Centromere: the constriction point of a chromosome to which spindle fibres are attached during cell division. Functions in the storage of genetic material.

Chapter 2

Activity 2.1:
a. Ee
b. Homozygous
c. 2 alleles

Activity 2.2:
a.

	B	b
b	Bb	bb
b	Bb	bb

Genotypes of offspring:
Bb and bb

b.

	B	b
B	BB	Bb
b	Bb	bb

50 per cent of possible genotypes would have the same Bb genotype as the parents. As both parents display the dominant phenotype (have one B allele), 75 per cent of possible offspring would also have the same phenotype (3 out of 4 have a dominant B allele).

Activity 2.3:

a. Mating of two heterozygous individuals:

	A	a
A	AA	Aa
a	Aa	aa

1 in 4 chance (25 per cent risk) of having an albino child.

b. Female carrier (Aa) x Albino male (aa):

	A	a
a	Aa	aa
a	Aa	aa

1 in 2 chance (50 per cent risk) of having an albino child (aa).
1 in 2 chance (50 per cent risk) of having a child with normal pigmentation, but a carrier for the albino allele.

c. 50 per cent risk of having a child who is albino like their father.

Activity 2.4:

This is an example of independent assortment as it is the only one of Mendel's principles that concerns the inheritance of two different genes.

Activity 2.5:

a. Pleiotropy. This example shows that the gene product contributes to different phenotypes.
b. Incompletely penetrant means that the gene in question is not always expressed in all individuals.

Chapter 3

Activity 3.1:

a. Both parents must be Bb genotype as both parents are unaffected but carriers of the recessive allele.
b. Carriers of a recessive allele only display the gene products of the dominant gene in their phenotype. Carriers are not affected by the single recessive allele.

Activity 3.2:

a. Autosomal dominant genes will always be expressed in the phenotype of the individual. Dominant genes can not be 'hidden' and passed through the generations in the same way as recessive alleles.
b. Risk is 75 per cent.
c. No, as the affected parent is homozygous recessive and can only pass down a dominant gene.

Activity 3.3:

a.

	C	C
S	SC	SC
C	CC	CC

CC = curly hair, SC = wavy hair

b. 1 in 2 chance (50 per cent risk) of child with wavy hair, 1 in 2 chance (50 per cent risk) of a child with curly hair.

c. No

d.

	S	C
S	SS	SC
C	SC	CC

e. Yes

f. Yes

Activity 3.4:

The blending of traits occurs in incomplete dominance.

Activity 3.5:

a. i) group B genotypes are: BB, BO
   ii) group O genotype is: OO

b.

	A	B
O	AO	BO
O	AO	BO

i) No
ii) No

c.

	B	b
b	Bb	bb
b	Bb	bb

i) Yes
ii) Yes

d. The biological father has type B blood as he is the only one that could possibly carry the recessive O allele.
e. This problem can be worked out by a process of elimination. Start off by working out who could be the parents of baby 2 who has type O blood. The only possibility here is parents number 1 as they are the only couple who could both pass a recessive O allele to their baby. Baby number 1 has type AB blood, which means that only parents number 3 could possibly pass on an A allele and a B allele. That leaves baby number 3 being the offspring of parents number 2.

Activity 3.6:

	A	a
A	AA	Aa
a	Aa	aa

a. Yes, genotype aa is possible.
b. Risk is 25 per cent of having a child with an AA genotype, which is lethal in this condition.
c. Risk is 50 per cent in this mating.

Chapter 4

Activity 4.1:
a. The genes in the pseudoautosomal region of the Y chromosome have corresponding alleles on the X chromosome. The genes in the male specific region do not have any corresponding alleles.
b. i) XX female
ii) XY female – there is no functioning SRY gene for the embryo to develop into a male.
iii) XXY male – due to the presence of the SRY gene on the Y chromosome, the embryo would develop into a male.

Activity 4.2:
a. More males than females would result in a sex ratio of greater than 1,000.
b. i) sex ratio is 832
ii) There are 83.2 males for every 100 females.

Activity 4.3:
War or conflict.

Activity 4.4:

John

	Xf	Y
X	XXf	XY
X	XXf	XY

Susan

f = Fabry disease

Results: all daughters would be carriers; none of the sons would be affected.

Activity 4.5:

a. males:	XY	normal vision
	XcY	colour blind
Females:	XX	normal vision
	XcX	carrier
	XcXc	colour blind

b. and c. colour blind parents must be XcXc and XcY

	Xc	Y
Xc	XcXc	XcY
Xc	XcXc	XcY

Results: all children would be colour blind as neither of the parents have a colour vision gene to pass on.

c. See answer b) above.
d. Yes, the daughter could have inherited a 'normal' vision allele from her mother and the colour blindness allele from her father. This would make her a carrier for colour blindness.
e. No. In order for the son to have full colour vision he would have to inherit a colour vision allele from his mother. His mother has two colour blindness alleles; therefore any son would be colour blind.
f. This woman must be a carrier of her father's colour blindness gene: her genotype would be XXc. If she mated with a colour blind male (genotype: XcY) then there is a 50 per cent chance that her son would be colour blind and a 50 per cent chance that her daughter would be colour blind.

Activity 4.6:
Possible genotype would be XXY.
This study found that the boys who developed Rett syndrome were either an XXY genotype or were genetically mosaic for this condition due to a mutation occurring in one of the cells during early embryonic development.

Activity 4.7:
As this gene is situated in the pseudoautosomal regions of the sex chromosomes, it behaves in the same way as an autosomal trait. It is therefore possible for father to son transmission of genetic material in the pseudoautosomal regions to take place. The risks are the same as in an autosomal dominant inheritance – 50 per cent.

Chapter 5

Activity 5.1:
a. ½ x ½ x ½ x ½ = 1/16
b. ½
c. Each child has a 1 in 4 chance of being affected. Probability that both will be affected is ¼ x ¼ = 1/16
d. Probability of Aa is ½, probability of BB is ¼, probability of Cc is ½.
   Answer is: ½ x ¼ x ½ = 1/16
e. i) ¾ x ¾ x ¾ x ¾ x ¾ = 243/1024
   ii) ¾ x ¾ x ¾ x ¾ x ¼ = 81/1024
   iii) ¾ x ¾ x ¾ x ¼ x ¼ = 27/1024
   iv) ¾ x ¾ x ¼ x ¼ x ¼ = 9/1024
   v) ¾ x ¼ x ¼ x ¼ x ¼ = 3/1024
   vi) ¼ x ¼ x ¼ x ¼ x ¼ = 1/1024

Activity 5.2:
Phenotypic traits that demonstrate measurable variation are termed quantitative. Only polygenic traits can be quantitative as monogenic traits are either present or not (do not demonstrate measurable variations).

Activity 5.3:
- Monogenic trait – a single trait encoded for by a single gene;
- Polygenic trait – a single trait encoded for by a number of related genes;
- Multifactorial trait – a genetic trait whose expression is affected or modulated by external factors.

Activity 5.4:
a. Heritability is an estimated measurement of how much the genes contribute towards the condition. Empiric risk is a measurement of the prevalence of the condition within a population, and the relationship of an individual to an affected individual. Both are population statistics.

b. The existence of different gene pools can result in different heritability rates between different populations.

Activity 5.5:

a. Definitions:
   i) Phenocopy: an environmentally-caused trait that is not genetic.
   ii) Pleiotropy: when a single gene is responsible for a number of distinct and apparently unrelated traits.
   iii) Penetrance: the proportion of individuals with a specific genotype who display the expected phenotype.
   iv) Heritability: the amount of phenotypic variation that can be ascribed to a genotypic variation.
b. Environmental factors such as increased calorific intake and lack of exercise have resulted in the tripling of obesity rates in the UK over the past 20 years. Genetic factors include susceptibility genes. As there has been a huge increase in just one generation it is clear that environmental factors have a large impact.

Chapter 6

Activity 6.1:

- replication errors in DNA;
- certain drugs and radiation.

Activity 6.2:

X chromosome. Monosomy X naturally occurs in all males as there is only one copy of the X chromosome. Monosomy X in females results in Turner's syndrome, which is compatible with life. Autosomal Monosomy is not compatible with life in humans.

Activity 6.3:

Familial Down's syndrome is an example of a balanced translocation.

Activity 6.4:

A numerical mutation results in a whole chromosome being added or deleted from the genome. A structural mutation means that varying lengths of an individual chromosome are added, moved or deleted; this does not affect the number of chromosomes present in the cell.

Activity 6.5:

a. This is an example of a point mutation.
b. This is a missense mutation as the original codon coded for the amino acid Serine, which has now been replaced by the amino acid Arginine. The replacement of AGC by AGA has resulted from a missense point mutation.

Chapter 7

Activity 7.1:

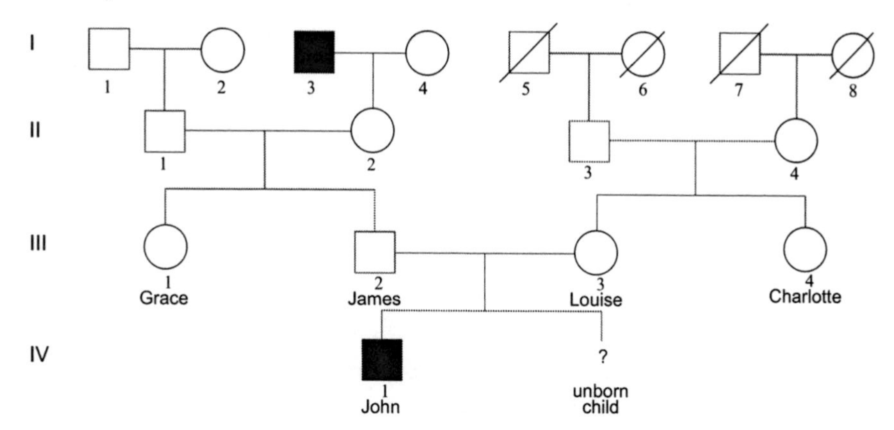

Activity 7.2:
III 2

Activity 7.3: Autosomal dominant.

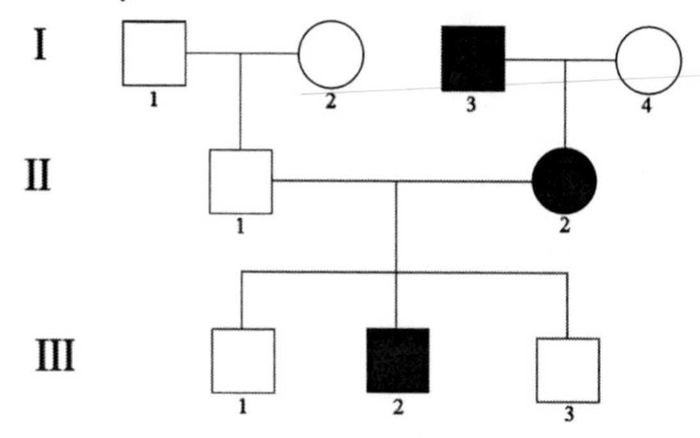

Activity 7.4:

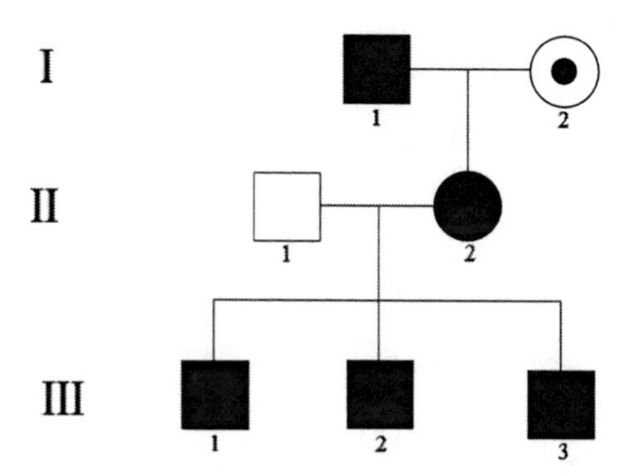

All sons would be colour blind as they only inherit their X chromosome from their mother. Both of Elizabeth's X chromosomes carry this gene, so all sons would be affected. Elizabeth's mother must have been a carrier for this gene (represented by a circle with a dot in the centre) as Elizabeth has inherited an X chromosome from each parent.

Chapter 8

Activity 8.1:
This is a reflective question.

Activity 8.2:
Problems may arise from false positive or false negative readings. Many conditions are due to the interaction of multiple genes and the environment so these tests cannot give accurate details for multifactorial conditions, only indicate the risk. A person who has been identified as high risk will not necessarily develop that condition.

Activity 8.3:
There are many arguments against germ line gene therapy due to the unknown long-term effects of the therapy. However, you might have your own views regarding both somatic and germ line gene therapy.

Activity 8.4:
- Cure rather than control of symptoms;
- Reducing risk of developing familial conditions;
- Reduced health care costs in the long term.

Activity 8.5:
- The short lived nature of gene therapy;
- The immune response;
- Problems with viral vectors;
- Multigene disorders like, for example, heart disease, Alzheimer's disease, arthritis and diabetes, are caused by the combined effects of variations in many genes.

Activity 8.6:
- What is considered 'normal'? Who decides this?
- Are disabilities diseases? Do they need to be cured or prevented?

Activity 8.7:
- More powerful drugs;
- Decreased cost of health care as a result of fewer adverse drug reactions, failed drug trials and an increase in the range of possible drug targets.

Activity 8.8:
- Right drug;
- Right dose;

- No adverse reactions;
- No trial by other drugs needed to find the correct drug and dose;
- Safer and more effective medicines.

Activity 8.9:
Other factors are just as important such as diet, weight, lifestyle and other medicines that the patient might be taking (polypharmacy). Testing out the genetic patterns with individual drug responses could be very time consuming and very expensive.

Chapter 9

Activity 9.1:
Proto-oncogenes are only active when the cells need to divide.

Activity 9.2:
Oncogenes act as dominant genes so, therefore, only one oncogene needs to be in a mutated oncogene form for a tumour to develop.

Activity 9.3:
Tumour-suppressor genes are always active. Tumours will only develop through a loss of function of a tumour-suppressor gene. Tumour-suppressor genes act as recessive genes; only through the loss of function of both genes will a tumour occur.

Activity 9.4:
You might have thought of a variety of carcinogenic materials. One example that you might have thought about is asbestos.

Activity 9.5:
Sun-beds use ultraviolet light that can alter DNA base linking.

Activity 9.6:
Vaccination is available against the human papillomavirus and Hepatitis B.

Activity 9.7:
Oncogenes are somatic mutations and are not passed on to any offspring. Tumour-suppressor genes can be inherited through germ line mutations and therefore can cause familial cancers.

Chapter 10

Activities 10.1 to 10.10: are reflective questions.

Activity 10.11: Paroxysmal nocturnal haemoglobinuria is an *acquired* genetic disorder, which is *not inherited*. There is no known risk of Alice's child inheriting this disorder.

GLOSSARY

Acrocentric chromosomes: have the centromere positioned so that one arm of the chromosome is much shorter than the other. Acrocentric chromosomes are chromosomes 13, 14, 15, 21 and 22 and are all able to take part in ***Robertsonian translocation***

Adenine: a DNA base.

Adult onset: a genetic condition that only manifests in adulthood.

Affected: an individual who displays the symptoms of a certain genetic condition.

Alleles: a pair of genes occupying corresponding sites on homologous chromosomes. Different forms of alleles produce variations in inherited characteristics such as eye colour and blood type.

Amino acid: the building block of protein.

Amniocentesis: a procedure that involves the removal of a small amount of amniotic fluid.

Aneuploidy: the occurrence of one (or more) extra or missing chromosomes.

Aniridia: the underdevelopment of the iris of the eye.

Antigen: any substance that causes the immune system to produce antibodies against it.

Apoptosis: the genetically controlled programme of cell death. This is the normal way of disposing of old or damaged cells.

Autonomy: an ethical principle – the right to make one's own decisions.

Autosomal dominant: genetic conditions that occur when a mutation is present in one copy of a gene situated on an autosomal chromosome.

Autosomal recessive: genetic conditions that occur only when a mutation is present in both copies of a given gene on an autosomal chromosome.

Autosome: any chromosome other than the sex chromosomes. Humans have 22 pairs of autosomes.

Base: part of the structure of DNA. Bases occur in pairs – adenine pairs with thymine and guanine pairs with cytosine.

Beneficence: an ethical principle – to do good.

Cancer: an abnormal new mass of tissue that is invasive and/or metastatic.

Carcinogen: a chemical or physical agent that causes cancer.

Carrier: an individual who does not express a recessive gene (heterozygote). As this recessive gene is not expressed in an individual's phenotype, the individual is said to be a carrier for that gene/trait.
Centromere: the constricted region of a chromosome that separates it into a long arm (q) and a short arm (p).
Chorionic villus sampling: a procedure that involves the removal of a small sample of cells from the placenta.
Chromatin fibre: the DNA and proteins that make up the chromosomes.
Chromosomal abnormality: a numerical or structural change in a chromosome.
Chromosome: a DNA molecule that is combined with proteins to form a linear structure.
Co-dominance: the expression of both alleles in a heterozygote.
Codon: a set of three DNA bases that codes for one amino acid.
Codon repeats: repetition of the same codon base.
Congenital: a condition that is present from birth.
Consanguinity: the genetic relatedness between two mating individuals who are descended from a common ancestor.
Consultand: an individual who consults a clinician for genetic advice.
Cytosine: a DNA base.
Degeneracy: different codons encoding for the same amino acid.
Deletion: a type of mutation that results from the loss of DNA from a chromosome.
***De novo* mutation**: a mutation that neither parent possessed.
Deoxyribonucleic acid: a biochemical structure that forms genes.
Diploid: having the full number of paired chromosomes (twice the haploid content). Most cells except the gametes have a diploid set of chromosomes.
DNA: see deoxyribonucleic acid.
Dominant: a gene that is always expressed.
Duplication: a type of mutation that involves the production of one or more copies of any piece of DNA.
Dynamic mutation: a progressive change in the DNA that affects the degree of expression within a gene.
Dysplasia: abnormality in the maturation of cells.
Embryo: a stage of development lasting eight weeks after the fertilisation of an egg in humans.
Empiric risk: the observed frequency of a disease in a given situation.
Enzyme: a protein that catalyses specific biochemical reactions.
Epicanthal fold: a skin fold of the upper eye lid covering the inner corner of the eye.
Epistasis: the interaction of different genes, where one gene masks the effect of another.
Eugenics: the study of methods of improving genetic qualities by selective breeding.
Ex vivo: occurring outside the body.
Familial: genetic traits that are transmitted through and expressed by members of a family.
Family history: the genetic relationships within a family together with their medical history.

First degree relative: parents, children or siblings of an individual.

Foetus: a stage of development lasting from nine weeks post fertilisation up to birth.

Frameshift mutation: a loss or gain of nucleotide bases by a number that is not divisible by three. This results in the reading frame of the DNA bases being out of synchronicity.

G0, G1, G2: phases within the mitotic cell cycle.

Gamete: a specialised reproductive cell which only has the haploid number of chromosomes.

Gene: the basic unit of inheritance.

Gene therapy: the correction of an inherited disorder at gene level.

Genetic counselling: a process of communication to aid at-risk or affected individuals in understanding the condition, risks, transmission and options available.

Genetic data: any form of information relating to both the phenotype and genotype of an individual.

Genetic linkage: different alleles that are closely positioned on the same chromosome.

Genetic screening: analysing a group of people to determine their genetic susceptibility to a particular condition.

Genetic testing: examination of genetic material to identify alterations.

Genetics: the branch of science that studies heredity.

Genome: all the genetic material held in the chromosomes.

Genotype: an individual's genetic make-up.

Germ line mutation: a mutation in every cell of the body. This mutation can be passed on to future generations.

Guanine: a DNA base.

Haemangioblastomas: benign blood vessel tumour of the eye or of the central nervous system.

Hamartoma: a benign tumour-like growth.

Haploid: a single set of unpaired chromosomes, as in a sperm or ovum.

Haploinsufficiency: a situation where a single normal allele is not able to encode for all the required amount of protein (normal amount is provided by two functioning alleles).

Haplosufficiency: a situation where only one functioning allele is able to encode for sufficient amount of protein.

Heredity: the transmission of genetic material from parent to offspring.

Heritability: the proportion of phenotypic variance attributed to variance in genotype.

Heterozygote: possessing two different forms of a particular gene, one inherited from each parent.

Heterozygous: having different alleles within the same gene.

Histone: a basic protein that forms the unit that DNA coils around.

Homologous: two representatives of each type of chromosome (a chromosomal pair).

Homozygote: possessing two identical forms of a particular gene, one inherited from each parent.

Homozygous: having identical alleles within the same gene.

Hypotonia: low muscle tone.

Imprinting: a semi-permanent modification of a gene that affects its expression. Imprinting can be changed in a subsequent generation.

In situ: 'in the place' – occurring at the actual site.

In vivo: 'in the living' – occurring within the body.

Incomplete dominance: being expressed or inherited as a semi-dominant gene or trait.

Informed consent: communication of information that enables an individual to achieve autonomous, informed decision making.

Inheritance: the transmission of genetic characteristics from parents to offspring.

Insertion: a type of mutation where genetic material from one chromosome is inserted into another.

Inversion: a type of mutation in which a segment of the chromosome breaks off and reattaches in the reverse direction.

Justice: an ethical principle – the fair and impartial treatment of all clients.

Karyograph: a photo micro-graph of chromosomes arranged in size order.

Karyotype: the chromosome complement within the cell nucleus.

Lethal allele: a mutated allele that results in the premature death of an individual.

Locus: the location of a specific gene on a chromosome.

Malignancy: the sustained proliferation of cells and the ability to invade other tissues.

Meiosis: cell division resulting in gametes that have half the normal number of chromosomes.

Mendel's Laws: a set of rules governing the inheritance of single gene traits as observed by Gregor Mendel.

Mendelian Inheritance: a pattern of inheritance displayed by a single gene trait that fits one of the standard patterns first described by Gregor Mendel.

Messenger RNA: RNA molecules that serve as templates for protein synthesis.

Metacentric chromosome: having the centromere positioned so that both arms of the chromosome are of equal length.

Microcephaly: a neurodevelopment disorder where the head circumference is much smaller than normal.

Missense mutation: a base mutation that results in coding for a different amino acid.

Mitochondrion (plural **Mitochondria**): a cellular organelle that supplies the cell with chemical energy.

Mitosis: cell division resulting in cells that have the normal number of chromosomes.

Mode of inheritance: the manner in which a genetic trait is transmitted from one generation to another.

Monogenic: a trait which is controlled by a single gene.

Monosomy: a human cell with only 45 chromosomes (one missing).

Mosaicism: two populations of cells with different genotypes being present in one individual.

Multifactorial inheritance: a pattern of inheritance caused by the interaction of one or more genes and the environment.

Mutagen: an environmental agent that causes a mutation.

Mutation: a permanent structural alteration in DNA.

Neonatal: the first four weeks of life.

Non-maleficence: an ethical principle – to do no harm.

Non-penetrance: when a genetic trait present in the genome is not expressed in the phenotype.

Nonsense mutation: a single DNA base substitution resulting in the termination of the protein structure.

Nucleosome: a complex of four paired histone molecules.

Nucleotide: compound of a DNA base linked to ribose or deoxyribose plus phosphoric acid.

Occiput: the back portion of the head.

Oncogene: a mutated proto-oncogene that results in uncontrolled, accelerated cell growth and cell division.

Organelle: a differentiated structure within a cell that performs a specific function.

p: the chromosomal short arm.

Pedigree: a pictorial representative of the relationships within a family.

Pedigree analysis: analysis of inheritance patterns within a pedigree.

Penetrance: the proportion of individuals of a specific genotype that display the expected phenotype.

Peptide bonds: the links within a chain of amino acids.

Personalised medicine: the systematic use of preventative, diagnostic and therapeutic interventions that use genome and family history to improve health outcomes.

Premutation: a situation where there is an expansion of chromosomal triplet repeats beyond the normal ranges, but insufficient to cause disease.

Pharmacogenetics: the study of genetic variation that gives rise to different drug responses.

Pharmacogenomics: the application of genetic knowledge to the development of new drugs.

Phenocopy: a phenotypic trait that resembles the trait expressed by a certain gene but in an individual who is not a carrier of that gene.

Phenotype: the visible or measurable properties of an individual resulting from the interaction of their genes and the environment.

Plasmid: double stranded DNA that forms a circle. Found in some viruses and used as a vector.

Pleiotropy: when a single gene is responsible for a number of distinct and often seemingly unrelated phenotypic traits.

Point mutation: the substitution of one base for another.

Polygenic genes: genes that vary in their expression due to the effect of other genes and environmental factors.

Polymorphism: a common variation in the sequence of DNA in humans.

Polypeptide: a chain of amino acids.

Pre-natal diagnosis: the biochemical, genetic or ultrasound tests performed during pregnancy to determine if the foetus is affected.

Pre-symptomatic testing: the genetic analysis of an unaffected individual who is at risk of developing a certain disorder.

Probability: the chance of inheriting certain genes or a genetic condition.

Proband: a family member with a genetically determined trait who first came to the attention of a clinician or investigator.

Proto-oncogene: a normal gene that controls cell division.

Pseudo-autosomal region: similar regions on the X and Y chromosomes where crossover occurs.

Punnet squares: probability diagrams that illustrate the possible genotype of offspring.

q: the chromosomal long arm.

Rare: a genetic disorder occurring in less than 1 in 5,000 births.

Recessive: a gene that is not expressed if paired with a dominant gene. A recessive gene is only expressed if both paired genes are in a recessive form.

Ribonucleic Acid (RNA): a biochemical structure that acts in transcription and translation of genetic information. Essential for protein synthesis.

Risk assessment: a quantitative or qualitative assessment of an individual's risk of carrying a certain gene mutation, or of developing a particular disorder.

RNA: see ribonucleic acid.

Robertsonian Translocation: the joining of two acrocentric chromosomes at the centromere, resulting in the loss of the short arms of the chromosomes.

Screening: see genetic screening.

Sex chromosomes: the two chromosomes that encode for gender.

Sex-limited trait: a trait that is only expressed in one sex, even though the trait might be an autosomal trait.

Sex limitation: sex-related expression of an autosomal gene.

Sex-linked: a gene that is located on the X chromosome.

Sex ratios: the ratio of males to females born within a population.

Silent mutation: a base mutation that results in encoding for the same amino acid.

Single gene disorder: a disorder resulting from a mutation in a single gene.

Somatic mutation: a mutation that has occurred in a non-sex cell. This type of mutation is not passed on to future generations.

Sporadic mutations: an alteration in DNA sequence that occurs at random.

Submetacentric chromosome: having the centromere situated so that one arm of the chromosome is slightly shorter than the other.

Susceptibility gene: a gene that confers predisposition to a disorder.

Syndrome: a set of phenotypic features that occur together as a characteristic of a disease.

Telocentric chromosome: having the centromere positioned right at the end of the chromosome (not present in humans).

Telomere: the DNA sequences located at the end of a chromosome.
Teratogen: an agent that causes birth defects.
Teratogenic: causing birth defects.
Therapy: see gene therapy.
Thymine: a DNA base.
Trait: a distinguishing characteristic.
Transcription: the process of synthesising messenger RNA from DNA.
Transfer RNA: RNA molecules that carry specific amino acids to a site specified by an RNA codon, for protein synthesis.
Translation: the process of synthesising an amino acid sequence (a protein) from messenger RNA.
Translocation: a type of mutation involving the transfer of DNA from one chromosome to another.
Trinucleotide repeat: a sequence of three nucleotides repeated on the same section of a chromosome.
Trisomy: a human cell where there are 47 chromosomes (one extra).
Tumour: an abnormal new mass of tissue.
Tumour-suppressor gene: a gene that codes for a protein that is involved in suppressing cell division and corrects DNA replication errors.
Unaffected: an individual who does not display any of the symptoms of a specific genetic condition within their family, or an individual who does not have the mutated gene in question in their genome.
Uracil: a nitrogenous base found in RNA but not in DNA. Capable of forming a base pair with adenine.
Vector: a plasmid that contains foreign DNA that is used in gene therapy.
X chromosome: a sex chromosome. Females have two X chromosomes, males have just the one.
X-linked inheritance: the inheritance of the genes situated on the X chromosome.
Y chromosome: the sex chromosome which is paired up with an X chromosome in males. Females do not have a Y chromosome.
Zygote: a diploid cell resulting from the fusion of two gametes.

INDEX

Added to a page number 'f' denotes a figure, 't' denotes a table and 'g' denotes glossary.